CATCH THAT TIGER

Noel Botham and Bruce Montague

CATCH THAT TIGER

CHURCHILL'S SECRET ORDER THAT LAUNCHED THE MOST ASTOUNDING
AND DANGEROUS MISSION OF WORLD WAR II

JOHN BLAKE

Published by John Blake Publishing Ltd,
3 Bramber Court, 2 Bramber Road,
London W14 9PB, England

www.johnblakepublishing.co.uk

www.facebook.com/Johnblakepub facebook
twitter.com/johnblakepub twitter

First published in hardback in 2012

ISBN: 9781857826609

British Library Cataloguing-in-Publication Data:

A catalogue record for this book is available from the British Library.

Design by www.envydesign.co.uk

Printed in Great Britain by CPI Group (UK) Ltd.

3 5 7 9 10 8 6 4 2

Papers used by John Blake Publishing are natural, recyclable products made from
wood grown in sustainable forests. The manufacturing processes conform to the
environmental regulations of the country of origin.

Every attempt has been made to contact the relevant copyright-holders,
but some were unobtainable. We would be grateful if the appropriate people
could contact us.

For Lesley and Barbara

ACKNOWLEDGEMENTS

The events described in this book are based upon the occasional journal of Major A.D. Lidderdale, which he noted in situ as the story unfolded.

During his lifetime, Douglas Lidderdale felt obliged to remain silent with regard to his mission to 'catch a Tiger'. In a letter to Captain James Henderson dated 4 November 1986, Doug wrote, 'I was instructed to tiptoe away from the subject [*of the Tiger tank*]! Which I did, yet with misgivings which have bothered me until today.'

In the same letter he wrote that, in his earlier response to Captain Henderson's request for information about the Tiger, 'I could not authorise that personally, but put your request to the Senior Security Officer of the Department of Tank Design at Chobham which shared the site with the Fighting Vehicle Proving Establishment and the School of Tank Technology. The immediate reaction was not to tell anyone anything...'

All the events described in this book actually occurred during

1942 and 1943. The authors have tried to be historically accurate in naming all the Allied fighting units involved in the various battles which are mentioned, and apologise if some which took part are not recorded here. All of the principal characters were real people but some of the minor characters have been amalgamated for the sake of clarity. Intimate conversations are of a speculative nature.

The authors would like to thank the staff at the National Archives who were so helpful and patient in dealing with their many complex military enquiries. Also thanks to David Willey, director of the Bovington Tank Museum in Dorset where Tiger 131 is now a star of their 300 tank exhibits. He and the archive staff at the Museum were extraordinarily helpful in unearthing some of the long-buried secrets that initially hampered our research, including some of Douglas Lidderdale's personal correspondence.

They are also indebted to military historians Major Peter Gudgin and Ronald Addyman for their splendid books on the 48th Battalion Royal Tank Regiment and the 51st Royal Tank Regiment, the Naval and Military Press, and the many authors, British, American and German, who have written about the Tunisian campaign and the Tiger tank.

PROLOGUE

Rock and Country star Dave Travis had always known his father, Colonel Douglas Lidderdale, was a hero. Ever since he was a small child, David had heard endless family stories of how the great Winston Churchill had sent Col. Lidderdale to Tunisia on a top-secret mission to capture a German Tiger tank.

But it was only after Col. Lidderdale's death in 1999, at the age of 86, that Dave – whose real name is David Lidderdale – began to slowly go through his father's papers, many of which were stamped 'secret' by the War Office.

A story of quite extraordinary heroism and daring was slowly confirmed. In a fierce, close-combat shootout, Col. Lidderdale had grabbed Germany's greatest war machine from the North African battlefield and brought it back to London, where a grateful Churchill proudly displayed it on Horse Guards Parade.

This was one of the great untold stories of the Second World War.

The unlikely event that brought about the writing of this

book was a casual lunch between singer and guitarist David, who has recorded more than 20 albums, and the authors. The three have been friends for more than 30 years.

It was the first time anyone, outside of the Lidderdale family, had heard the story of how his father successfully led one of the most daring and incredible secret missions of the war on the personal order of England's greatest wartime leader. *Catch That Tiger* is the result of that remarkable lunchtime revelation.

CHAPTER ONE
20 APRIL 1942

At eleven o'clock on Monday morning, the most hated man in the world woke alone and naked in his pristine, almost sterile bedroom in Wolf's Lair, East Prussia.

Adolf Hitler, the megalomaniac, who in 1942 ruled most of Europe with an iron will that he believed would one day determine the fate of the entire world, reached for his spectacles and screamed for his valet.

Sturmbannführer Heinz Linge, who had been waiting outside the door for several minutes, knocked once and entered carrying a tray on which was laid the Fuhrer's regular frugal breakfast of tea, biscuits and an apple. A square white envelope was tucked against the teapot. Heinz stopped by the bedside, snapped the heels of his boots together and flung his right arm forward in the Nazi salute favoured by his master.

Hitler waved his hand languidly as he fixed Heinz with a stare of eye-bulging expectancy.

'Did he do it?'

'Sir?'

'Did Rommel get me my birthday present?'

Heinz placed the tray on the table by the bedside and gave an uncertain glance through the window. He had no intention of spoiling his Fuhrer's birthday surprise.

Rain spattered against the windows and a harsh wind rattled their frames. Hitler slammed his hand down on the bed covers.

'Tobruk, you imbecile. Rommel promised me he would take Tobruk by my birthday. What's the news?'

'I'm afraid there has been no news as yet, Mein Fuhrer,' said Heinz, 'Field Marshal Rommel said last night his men are at the gates of the city. They have advanced 300 miles from Benghazi taking everything in their path. The Field Marshal is anxious to fly back to Libya to take personal charge. He is here for your birthday, Mein Fuhrer.'

'But he promised,' Hitler began petulantly. Then he stopped himself. 'No matter. He can give me the news himself later. Now what is this?'

He picked up the envelope and took out a card with an intricately decorated swastika on its cover.

Heinz clicked his heels again.

'Happy Birthday, Mein Fuhrer.'

The ghost of a smile flickered across Hitler's usually impassive face. 'Well, not everyone forgot. How old are you, Heinz?'

'I was 29 last month, sir,' said Heinz, pouring the tea.

'I'm ahead of you by nearly a quarter of a century,' Hitler sighed.

There was a knock at the door. Hitler barked a harsh 'come in'.

The door opened and the voice of an untrained contralto started to sing 'Happy Birthday', and finished with 'Happy Birthday, dear Fuhrer, Happy Birthday to you'.

The singer was a woman of average height with sturdy

thighs and an ornament pinned into her freshly permed, light-brown hair that was cut just short of shoulder length. She carried a small birthday cake on which there were three flickering candles.

'Eva!' said Hitler, a rare smile broadening his thin lips, 'When did you get to Wolfsschanze?'

'This morning,' said Eva, giving her lover a coquettish look. 'Now, my dear Adi, you must make your birthday wish … and receive your gift.'

Hitler made a great fuss of blowing out the candles, then turned to his aide. 'Leave us, Heinz.'

Heinz gave the slightest acknowledgement and left the room. He closed the door and caught a glimpse of Eva Braun's naked legs as she slipped out of her knickers while clambering on to the bed, where Hitler was already throwing back the bedclothes.

Wolf's Lair was the headquarters of Hitler's Eastern operations. It had been built in 1941 for Operation Barbarossa, Hitler's grand plan for the invasion of the Soviet Union. Over 2,000 people worked at Wolfsschanze, so called because, in Old High German, 'Adolf' translates into 'noble wolf'. Zone One had ten vast bunkers, each protected by two-metre-thick steel-reinforced concrete. Zone Two contained the military barracks and compounds for Hitler's special security troops.

East Prussia had suffered its harshest winter for over a hundred years. Though it was now April, an icy wind still blew into the little town of Rastenburg, where Wolf's Lair was situated. All those gathered outside in the early afternoon to celebrate Hitler's 53rd birthday were shivering. A greatcoat covered Hitler and his uniform as he strode towards the marshalling yards bordering the Masurian woods. He was surrounded by members of his inner circle; Hermann Goering, the portly head of the Luftwaffe; Field Marshal Rommel, Head of the Panzer Division of Tanks; Martin Bormann, Hitler's chief aide; the newly appointed Minister for Armaments, Albert Speers; the War Minister, Wilhelm Keitel; and the chief of the OperationsStaff of OKW (German High Command), Alfred Jodl.

Running ahead of Hitler, and attempting to be as inconspicuous as possible, were two official photographers. One took photos on a stills camera, the other shot with 16mm movie film – both were recording every important movement of their beloved Fuhrer. Posterity was paramount.

Hitler was in high spirits. A birthday 'surprise' had been prepared for him, but the Fuhrer knew what was coming, and he looked forward to it. He sang in fractured English, the words of a popular song:

'You'd be zo nice to come home to.
You'd be zo nice by ze fire,
Ba ba zo nice,
Ba ba paradise
To come to and love.'

'Very nice,' said Reichsmarshall Hermann Goering, applauding. 'Cole Porter, isn't it?'

'One of his latest. Eva gave me the sheet music for my birthday. I am older than him, you know, by two years. I'll play it for you later.'

'I can hardly wait,' exclaimed Goering, stretching his jovial jowls.

Hitler claimed to have been taught to play the piano by conductor August Kubizek and later by gifted pianist and business friend Ernst Sedgwick Hanfstaengl, but those who heard the Fuhrer sight-read had their doubts.

Hitler came to an abrupt halt. His entourage shuffled forward and looked at him expectantly.

'Well, where is it?'

'Where is what, Mein Fuhrer?' asked Albert Speers innocently.

'My surprise; my new tank.'

Jodl pointed towards the woods. Seconds later, two huge tanks materialised and rolled slowly towards them on parallel paths.

'The one on the left is the Porsche model.'

'Good for Porsche,' said Hitler, obviously in some awe. Even his expectations were exceeded. 'It looks enormous.'

'Forty-five tons,' said Goering.

'What is its speed?'

'About 2mph.'

'Is that fast enough?'

'It's fast enough if it doesn't stop.'

As Goering spoke, the Porsche stopped abruptly and a plume of smoke rose from the command hatch on the turret.

'What's wrong with it?' asked the Fuhrer impatiently.

'I think it's broken down.'

Hitler looked crestfallen; like a child who had lost his toy. 'What about the other one?'

'That's the Herschel model. It's even heavier.'

The Herschel trundled relentlessly forward.

Hitler brightened. 'It looks bigger than our old Panzers.'

Rommel stepped forward and spoke softly. 'It's twice as big as anything we've had before.'

'Bigger than a Panzer?'

'And fiercer. They call it a Tiger.'

Hitler's eyes lit up. 'I like that name. Is it heavy?

'About 60 tons.'

Hitler blinked at that news. He could barely contain his excitement. 'It has a very big cannon.'

Jodl read aloud from the manufacturer's specifications. 'Fitted with an 88mm cannon, accurate to 2,000 yards. Carries 90 shells that weigh nearly a ton, and its steel armour casing is 120mm thick.'

'Is that better than the Russians?'

'Twice as good as the Russians, three times as good as the Americans and four times better than the British.'

Hitler's eyes glistened. He was breathing faster. It did not take much to capture his imagination. 'How soon can we have them?'

'As soon as you confirm the order.'

'I want them now. I plan to smash our way through to Stalin's hideaway and we need some for you too, my dear Rommel, so you can take over all of North Africa.'

Rommel clicked his heels. 'As you say, Mein Fuhrer.'

Hitler raised his hand. 'One more thing...'

'Mein Fuhrer?' said Jodl.

'What is that big tube sticking up at the back of the Tiger?'

'A snorkel, sir.'

'A snorkel! I thought snorkels were on U-boats.'

'True, sir. But these tanks can go underwater.'

Hitler looked at Jodl in disbelief.

Speers interceded. 'We have solved the problem of crossing rivers after the enemy has fled and blown up the bridges behind them. We no longer need to waste valuable time rebuilding the bridges. These tanks – Tigers as you call them – will cross rivers underwater. They can submerge to a depth of 13 feet.'

Hitler gave a short laugh – an unusual occurrence. 'Then nothing can stop us. We can go anywhere?'

'Yes, Mein Fuhrer.'

Hitler slapped Speers on the back. 'You have given me the most wonderful birthday present. Thank you. Thank you.'

He turned in a semicircle and raised his hand in the familiar Nazi salute. 'I thank you, gentlemen. Today you have won the war.'

Hitler put an arm on Rommel's shoulder and marched with him towards the tank. 'Erwin, now I want to see inside. In one of these you can drive underwater to Tobruk from Egypt. As soon as you have kept your promise and taken it, that is.'

As the top brass of the German High Command inspected the prototype Tiger tank, the photographers continued to capture every moment.

CHAPTER TWO

AUGUST 1942

Operation Barbarossa had been a disastrous enterprise for the Germans. The fierce winter that ushered in 1942 brought death and destruction to Germans and Russians alike on an unprecedented scale. By the summer of that year, Hitler was determined to press on regardless, and he issued Fuhrer Directive No. 41 to order yet another offensive against the oil-rich region of Baku. This was to become known by the Allies as Operation Case Blue.

With a million Soviet soldiers killed and countless thousands captured, many young Russian men were eager volunteers to escape the carnage, if only briefly, and attend weapons-training courses in the country that had now become a reluctant ally – England.

Major Aubrey Douglas Lidderdale's spirits were at a low ebb. The Corps of the Royal Electrical and Mechanical Engineers (REME) was being created, and Doug was to be one of its founding officers. He knew the motto was going to be '*Arte et Marte*' – 'by skill and fighting'. Doug was looking forward to it all, but, until 1 October he was still an engineer in the Royal Army Ordnance Corps (RAOC) and not a happy man.

Having designed the first Leyland twin-engined tank unit, he had signed up immediately after the outbreak of war and was commissioned within months. It took another year before the High Command realised they had a highly qualified engineer in their midst, but, as soon as his worth became apparent, they rapidly promoted him to Major and put him in charge of the newly formed Tank Training unit. Britain had invented the tank, thanks to Churchill, but enemies and allies alike had woefully outdistanced her in the arms race. It had taken an all-out war for the High Command to try to catch up. Thousands of new drivers and maintenance crews were being recruited and they had to be trained. This was 28-year-old Major Lidderdale's job.

He had devised a whole raft of courses on how to drive and maintain tanks. Now he was in charge of a British army tank instruction school in West Kent and his current crop of Russian trainees was worse than bloody useless. Doug confided to one of his driver-instructors, Corporal Bill Rider, that the word 'crash course' took on a peculiarly appropriate meaning when applied to some of his current students – a platoon of disaffected Russians. This was not a promising bunch. The Russians seldom shaved, nor, Doug suspected, washed regularly either. They made no attempt to learn English and instructing them through an interpreter was tiring work.

The group had been brought to this desolate part of Kent and foisted on Doug by the powers-that-be in order to become experts in the art of tank warfare. However, as time went by, his success rate with them remained at precisely zero. Doug's despair grew visibly. Whatever he instructed them to do, there seemed to be a wilful determination on the part of the Russians to do the exact opposite.

The West Kent Country Club, which was only a mile away, had a well-stocked bar. For Doug, the club was growing more hypnotically attractive as the long, hot summer days wore on.

Following the group's latest tests, which they had managed to fail spectacularly for the third time, Doug succumbed to temptation and treated himself to a spiritual reviver at the Country Club bar. As the amber liquid began its soothing work, he resolved to confront the Russian Commandant and get to the bottom of this batch's perverse attitude once and for all.

He drove out of the club in 'The Beast' – the name he had given to the mongrel, two-seater sports car that he had spent many greasy hours building. Its unique appearance vaguely resembled a truncated road version of Malcolm Campbell's 'Bluebird' racer. Though not designed for the racing circuits, it was fast, with a Rolls-Royce engine tucked under the bonnet. Doug loved to drive it, but was sadly reconciled to the fact that, thanks to the latest petrol-rationing squeeze, he would not be able to keep The Beast on the road for much longer.

He was sadly contemplating this inevitability when he spotted a Tin Lizzie coming towards him. The old-fashioned Model T Ford had a young woman with long blonde hair at the wheel. As the two cars approached one another, the Ford spluttered to a halt.

Doug leaped from The Beast and removed his officer's cap. The blonde stepped out. She was tall and poised and dressed for an afternoon at the Country Club. Doug appraised her appreciatively. She had a lithe figure and the sort of legs that seemed to go on forever.

'Can I help?' Doug volunteered.

Man and woman looked at each other and liked what they saw. So struck was he, in fact, that for quite a moment he stood staring at her goofily. It was what the French call a *coup de foudre*, a lightning bolt to the heart.

After a few moments, the silence was broken.

'Do you know anything about engines?' asked the young lady.

Doug had to shake his head as if to remind himself what engines were. 'A little.'

'Well, I've just filled up with petrol, so it must be the engine.'

'Can I take a look?'

'Oh, please.'

'I thought they'd stopped building Model T Fords years ago.'

'Officially they did. But they still make them to order for you. At least they did until 1941.'

'It must be difficult to get replacement engines.'

'I think that's why they finally stopped.'

'If you open the bonnet, I'll take a gander.'

'That would be absolutely darling of you.'

Doug beamed. Being a darling to this lady would be the cat's pyjamas as far as he was concerned.

Ten minutes later, Doug emerged from the open bonnet covered with oily smudges.

'You had a dirty magneto,' he said.

She looked shocked. 'Dear me,' she said. 'I never even knew I had a magneto; let alone a dirty one.' She came towards him, took out a dainty handkerchief from her purse and rubbed at the smudges on his face.

Doug remained as immobile as a statue. He could smell her gentle perfume – Chanel. This was pure heaven.

'Have you got the starting handle?'

'Oh, no. This has got a self-starter.'

'Really? Oh, of course – custom-built, you said. Well, give it a bash. It might tick over now.'

The lady slid her long legs elegantly over the driving seat and switched on the ignition. The engine burst into life.

'Oh, that's wonderful,' she said. 'How can I ever thank you?'

'Well, you could start by telling me your name,' said Doug.

'Kathleen Crane,' she said. 'And you, Major?'

'Douglas. The chaps call me Doug.'

'Oh, dear, I couldn't call you Doug. I shall always call you Douglas.'

There was the tiniest pause.

'Always?'

The promise hinted at in this one word sent Doug's pulse racing.

She smiled at him. 'I'm often at the Country Club. That is, if I'm not nursing or dancing. Why don't you look me up sometime?'

'I rather think I will.'

'I look forward to it.'

'Where do you like to dance, by the way?'

'I dance where they send me.'

Doug must have looked puzzled because she sailed straight on. 'You see, I dance professionally. Not as often as I used to, but whenever I can get the engagements. Times are tough.'

'Oh, right.' A tongue-tied Doug stared at her with a daft, rictus smile on his face. His mind whirled. Here is a beautiful girl, he mused, and I want to see her again. A nurse who also happens to be a professional dancer. Oh, boy! If she says she owns a pub, I think this whole encounter might be part of some wonderful dream…

'What sort of car are you driving?' she asked, snapping Doug from his reverie.

'She's a hybrid I built myself. I call her The Beast.'

'I'd love to have a go in her.'

Doug nodded. 'You're on,' he said. 'I'll give you a go in her whenever you say the word.'

She waved farewell and drove on towards the club. Doug was beaming as he continued his journey to see the Russian Commandant.

A short time later, sitting at the mess bar, Doug found it difficult to concentrate on his Russian counterpart, Alexei, who was standing beside him nursing a large vodka.

Doug gathered himself and began. 'Your men have failed their tests three times,' he said. 'And each time they're getting worse. What on earth is going on? Do you think we're involved in some kind of board game? This is war, damn it, and if your blokes can't take it seriously then you can get your arses back to Russia. I'm wasting time on you that I just can't afford.'

Alexei reacted with tight-lipped reticence. It was not that he disliked Doug, but, like with Stalin, his boss, he had an innate distrust of Englishmen. His platoon was due to move back to Russia shortly, anyway, so maybe he could risk breaking a confidence to this usually gentle man with the penetrating grey-blue eyes. He sipped his vodka before speaking.

'We are men of Russian Revolution, you understand?'

'Yes, I understand that.' Doug sipped his whisky.

'You are English. You do not approve of Communism.'

'I don't approve or disapprove. What you do in your own country has nothing to do with me.'

'Before we leave Russia, we are all brought before the Commissar.'

'Oh, yes?' Doug could not see where this conversation was leading.

'The Commissar tells us that Britishers hate us Russians. He says you are only fighting on our side to defend yourselves. He says you will tell us everything wrong with these tanks. So that, when we go back to war, we blow ourselves up.'

'Your boss told you that?' Doug looked incredulous, downed his scotch and summoned the bar steward to refill both their glasses.

'You deny?'

'Certainly, I deny it.'

'OK. *Zhal*. But that is why my men listen to you carefully and then do opposite. Now you understand, yes? I will get them to apologise and they will do much better.'

Doug pondered on this. He did not really understand at all. All of a sudden, he could no longer be bothered to try.

He turned to his companion with a brighter face. 'Alexei, let's forget the whole thing and get drunk. Do you know what happened to me today?'

The Commandant shook his head and waited.

Doug laughed like a schoolboy. He had the faraway look of a man who had come face-to-face with a creature from outer space. 'I met the girl of my dreams.'

Alexei muttered a sentence in Russian, and Doug knew just enough of the language to realise that the other man envied him.

CHAPTER THREE

DECEMBER 1942

It was not until the early winter of 1942 that the new Tiger tanks rolled off the production lines. When they opened fire in battle, their long-range accuracy and the calibre of their weaponry inflicted horrific damage on the Allied forces.

Earlier, when Winston Churchill had been shown a photograph of the prototype Tiger with Hitler standing beside it, he had been convinced he was being fed disinformation. He could not conceive of a tank being so huge.

'Where did this photograph come from?' Churchill asked the security officer who had brought it to his attention.

'One of our agents in the field, sir.'

'Why is it so crumpled?'

'It was retrieved from a wastepaper basket, sir.'

'Who brought it in?'

'It was couriered by one of our moles, sir.'

'Can we trust him?'

'The agent is dead, sir.'

'I won't believe they've built a tank of this size until I see it myself.' Churchill's dismissiveness resulted in no serious attempt being made to obtain specifications of the new tank until the following year, by which time the Tigers had destroyed hundreds of Churchill and Sherman tanks and decimated the Russian tank brigades.

Once Hitler started sending Tiger tanks into battle in North Africa, it was not long before the gloom deepened in the Cabinet War Rooms, under the streets of Whitehall, where the heart of government functioned in relative safety away from the bombing above.

In the Map Room, the advances and retreats of the armies were continually updated. The Germans had blockaded the Western approaches to the British Isles with the intention of keeping vital supplies such as food and fuel from getting through. The progress of Allied convoys battling to keep the nation supplied had been slowed to a snail's pace.

Four duty officers sat close to their telephones, waiting to spring into action at a moment's notice. Winston Churchill came into the room, his head bowed deep in thought. Prone to insomnia, he was not in a buoyant mood. He held a narrow brass-topped cane in his left hand and an unlit cigar in the other. In the winter months of late 1942 and early 1943, a decisive change in the fortunes of war was proving elusive for the Allies.

The duty officers, taken unawares by the arrival, were slow to start rising, but Churchill gestured at them to remain seated. For a while, nothing could be heard but the ticking of a wall clock.

Eventually, Churchill spoke. 'It's very quiet.'

'Yes, sir.'

'Ominously so.'

'It's one o'clock in the morning, sir.'

'Wars don't shut down at night.'

None of the men responded. A red light flashed on one of the phones. The Lieutenant manning it answered immediately and gave the codeword. He listened for a full minute before replacing the receiver.

Churchill looked at him. 'Well?'

'The Fourth Infantry Division, sir.'

Churchill nodded. 'North Africa,' he said. 'Part of our First Army.' He possessed an almost photographic memory of where every battle unit was or, at any rate, where it should be. He gave a mischievous grin. 'Tell me something to make me feel ten years younger. Please tell me they've got Rommel.'

The Lieutenant shook his head. 'I'm sorry, sir. We've suffered heavy casualties.' He consulted his shorthand notes. 'We've lost thirty-eight tanks in the last two days.'

Churchill's face was impassive. He paused for thought. 'What sort of tanks were they? Churchills?'

'I'm afraid so, sir. Knocked out by the new German super tanks.'

'The Tigers, eh? We've got to get our hands on one of those things. Then our boffins can open one up and see how the box of tricks works. Until we do, we cannot turn the tables on them.'

The duty officers nodded in agreement.

Churchill was in one of his brown moods. 'Dear God! I hope most of the crews escaped. It's difficult to imagine a worse way to die than to be roasted alive inside a tank.

'They never mention being burned to death, you know. Call it being brewed up.'

Churchill paced restlessly for a few moments, his inner turmoil obvious. Suddenly, he stopped. 'We must get hold of an engineer – a tank expert – someone with the knowledge to catch us one of these damnable German Tigers, and smuggle it home.'

'It's a big thing to smuggle, sir,' said the Lieutenant. 'I've heard they weigh over 60 tons.'

'I've heard that rumour too. I've even seen a picture of the damn thing. We've got to have one. I'll send a specialist team to Africa. I want a list of our top tank engineers before reveille. Carry on, men.'

Churchill turned on his heel and headed for his own office.

In the corridor, Sir Charles Wilson, Churchill's personal doctor, confronted him.

Churchill tried to brush him aside. 'What the devil do you want, Wilson?'

Sir Charles didn't stand for any nonsense from his difficult patient. 'You are flying to Casablanca on Monday. You must get some sleep.'

'Stop fussing, man.'

'I insist that you sleep, sir.'

'You what? You are telling me what to do?'

'Yes, I am.'

Churchill turned and raised his voice. 'Thompson! Come here. I'm being kidnapped.'

Silently, out of the Prime Minister's private office the tall shape of Churchill's personal bodyguard, the taciturn Inspector Thompson, materialised.

'Yes, sir?'

'Wilson is manhandling me.'

This was a recurring situation with which all three men had grown familiar.

Sir Charles spoke quietly. 'Help me get this patient to bed.'

Churchill glared at them. These were men he had personally appointed and for whom he had the highest regard.

'Damn you both. I'll go quietly. But I'll sleep upstairs in my own bed. Give me one of those Phenobarbital things.'

The following year, Sir Charles Wilson would be created a Peer of the Realm.

CHAPTER FOUR
8 JANUARY 1943 – 1400 HOURS

In Germany, Tiger tanks were rolling off the Herschel production line at the rate of two a day. Douglas Lidderdale had read about them in the newspapers. They sounded formidable.

The second Friday in January would prove to be a fortuitous day for Doug, though he had no idea what fate had in store for him as he sat, unhappy and freezing, in the back seat of a staff car heading at great speed for London.

Beside him sat Major Desmond Morton, an officer who had been severely wounded in the First World War; the bullet was so close to his heart that it was too dangerous to remove. Morton was whisked away from the front line and appointed as one of the aides-de-camp to Field Marshal Haig. Now, 25 years later, at the age of 53, he had become Winston Churchill's personal intelligence adviser. He had a broad forehead, a neat moustache and slightly hooded eyes that seemed to penetrate into everything they saw. He wore a suit of the finest quality.

Owing to the rationing of petrol, there was little traffic on the roads. The car's wheels screeched as the blue-uniformed female driver took a bend just a little too fast.

Doug could contain himself no longer. 'What's the hurry? Can't you at least tell me where we are going, sir?'

Major Morton gave his moustache a protective rub with his forefinger. 'Top secret, Major,' he mumbled, barely opening his mouth.

'Are you aware I'm about to go on compassionate leave?'

'Indeed?'

'It's all square and above board. I'm set to get married on the 19th. To a dancer no less. One of Cochrane's young ladies.'

'Congratulations. I hope you'll be very happy.'

'Oh I will. But my fiancée is worried sick something might force us to cancel our wedding.'

'Don't worry, my dear chap. You'll have time to marry her before you go abroad.'

Morton clammed up. Doug wasn't certain whether the older man had let this information slip out accidentally or by design. He rather fancied the old warrior never said anything without meaning to.

He turned to face Morton. 'Abroad? Am I being posted abroad?'

'I can't possibly say, Major. There is a war going on. A lot of soldiers do find themselves abroad.'

'You know something, don't you?'

'I'm merely a liaison officer. I obey orders.'

'Who gave you orders to drag me away just as I'm about to get married?'

'I'm afraid I can't divulge that information.'

'Why not?'

'It would compromise security.'

Doug shook his head in disbelief. 'Look, all I do is train

20

people in the fundamentals of making tanks work at an Army Tank workshop.'

'I've read your file, Major. You were enlisted first into the RAOC. Your basic training was at Hilsea Barracks near Portsmouth.' Desmond Morton had done his homework. 'Then you were a founder officer of the Royal Electrical and Mechanical Engineers. Three months ago, your unit moved to King's Lynn to link up with 25th Tank Brigade. You train people to operate tanks.'

Doug nodded. 'I'm not really a tank man. After Charterhouse, I trained at Leyland Motors and qualified as a Chartered Engineer.'

Morton already knew this too. 'F.I. Mech.E and F.I. Prod.E. First-class degrees. Then you worked on the first twin-powered Matilda tank.'

'I also designed buses,' said Doug defensively. 'I set up the trolleybus service in Canton in China. But I'm not a clippie.'

'Quite so, Major, I really am fully briefed. If I may say so, I imagine the War Chiefs are more interested in your expertise with tanks than with trolleybuses.'

'Where are we going? Who am I supposed to see?'

Morton squinted out of the window. His face betrayed nothing. He said quietly, 'Charterhouse, eh? I went to Eton myself.'

'You're not going to let me in on anything, are you?' The tone of Doug's voice betrayed his annoyance.

'Careless talk costs lives,' replied Morton imperturbably.

Dusk fell as the car drove without lights into central London. The houses along the streets of Westminster looked curiously vulnerable with their iron railings removed – ostensibly to be taken to foundries and used for the war effort. The barbed wire, which had replaced the railings, rolled out in all directions. The strict blackout restrictions lent a sepulchral air to the cold streets of London.

The night was shrouded in black cloud. From his invisible eyrie, 170 feet above them, Horatio Nelson, the saviour of the British nation just 137 years previously, stared blindly towards the sea.

Finally, the car stopped in King Charles Street. Major Morton exchanged salutes with the two Marine Sergeants on guard at the St James's Park entrance.

'I am escorting Major Lidderdale,' said Morton. 'Here is my authority from Captain Adams.'

After inspecting the documentation, the sentries saluted and let the men pass through the large oak door. Inside, the two Majors made their way down a spiral staircase to what might once have been a huge wine cellar or an ancient tunnel built for drainage. At the bottom of the stairs, rigid as a Madame Tussauds waxwork, another armed Marine stood guard, his rifle bayoneted. The visitors nodded to him as they passed, then disappeared into a labyrinth of passages.

On either side were green-painted doors with numbers on them. Eventually, they reached Room 60 and stopped to salute a Marine Captain who checked their credentials. The lighting was stark – a combination of poorly concealed naked bulbs and yellowish strips. A young business-like woman in a smart Women's Auxiliary Air Force (WAAF) officer's uniform appeared and escorted the men into an anteroom. The WAAF was the women's section of the RAF. It had been formed before the war to do mundane tasks in order to free up men for active service. However, when war was declared, the WAAFs soon proved themselves to be indispensable.

Doug and Desmond Morton waited for another hour. In this troglodyte world, looking at the clocks was the only way of knowing whether it was day or night. Metallic-smelling air was piped in from the rotundas above. There was the constant chattering of a teleprinter in the vicinity and, somewhere, Gestetner copying machines thrummed and buzzed infuriatingly. The claustrophobic effect and the all-pervasive hum reminded Doug of being in one of his own tanks.

Finally, the brisk clatter of a lady's heels signified that they had not been entirely forgotten. A young woman in a neat black dress and a 'sensible' jumper approached and offered a welcoming smile. Joan Bright was the head of the Secretariat in the Cabinet War Rooms and was known for her willingness to go anywhere and meet anybody.

'If you would follow me, please, gentlemen.'

The two men got to their feet and found they had to trot to keep up with her.

As she walked, she talked. 'Mr Martin has asked me to say please don't keep the Prime Minister talking for longer than is absolutely necessary. He has an exhausting schedule on his agenda.'

'Who is Mr Martin?' enquired Doug.

'Mr Churchill's private secretary.'

She walked them through Room 66B and showed them into the smoky fug of Churchill's private office.

The room was covered in maps and books. Churchill sat behind a large desk on which lay a mountain of paperwork and three telephones. One of the phones was red and Doug assumed that this must be the Prime Minister's direct line to somebody important. President Roosevelt in the United States perhaps. Or even the King. At the Prime Minister's right elbow there was a large, half-filled brandy snifter. There were two ladies in the room sitting at smaller desks. Though the typewriters had been specially muted, the men still had to raise their voices above the occasional sound of typing.

Major Lidderdale came to a smart halt and saluted. Churchill acknowledged him and gestured for them to sit.

'Is this your man, Morton?'

'He seems perfectly qualified, sir.'

While Churchill shuffled the papers in front of him, Doug studied the famous man, noting the sparkle in his eyes and his ebullient mood. For a man of 68 with a reputation for over-indulgence and ill health, Churchill looked tired but in good spirits. Doug prayed that he himself would be as lively in 40 years' time.

Churchill looked up. 'Mr Lidderdale?'

'Sir.'

'I've heard a lot about you. Pretty genned up on tanks, are you not?'

'As an engineer, sir. I'm not really a military man.'

'We are all military men at the moment, Major.'

'Yes, sir.'

'I know something about tanks myself. In the Great War, when I was First Lord of the Admiralty, I ordered the first tanks to be built. And why, I can hear you asking, was a Sea Lord concerning himself with tanks? Well, I'll tell you, sir. At that particular moment, I didn't think of them as tanks. In fact, I called them landships. They were only called tanks to confuse the enemy in the event of their managing to intercept our requisitions. I trusted that they may have been foolish enough to assume we were building water-storage facilities.'

Churchill chuckled and looked at his cigar, which had gone out.

'In the trenches, hundreds of thousands of our brave young men were being needlessly slaughtered – going over the top, and getting nowhere. My God! When I think of it.' Churchill paused for a moment. 'The sole reason for pressing ahead and getting the tanks built was to overcome the stalemate. A tank can get over the top and through the barbed wire and force an advance.'

Again, Churchill stared reflectively at the end of his dead cigar. 'Do you know why Ludendorff ended the Great War?'

Doug looked puzzled. 'I had always supposed it was because we had defeated him, sir.'

Churchill sniffed. 'It was because of the tanks. On 8 August 1918, I attacked the German lines on the Western Front with 430 tanks. Our advance overwhelmed the Hun. They knew then that they had no choice but to sign the Armistice. So you see, I know something about tanks.'

'We call the present generation of tanks Churchills in your honour, sir,' said Doug, hoping not to sound too mundane.

Churchill peered at Doug over his glasses. 'The Churchills have the advantage of being able to traverse hilly ground with some ease and rapidity. But the Germans' new Tiger tanks, although more cumbersome, and no good at all on hills, are deadly accurate and almost impregnable. We did destroy one a month ago but we demolished it so thoroughly that, apart from its dimensions, we have no real knowledge of how it is put together. What I need is a complete Tiger.'

'As a matter of interest, sir, what are its dimensions?'

'It's 27 feet from muzzle to aft and about 12 feet wide. It weighs around 60 tons.'

Doug let out a low whistle.

Churchill grinned and said, 'Quite so. I am told that the Tiger's cannon fires an 88mm round with deadly accuracy. Neither we, nor our Allies, have anything nearly so powerful on our tanks. We have to know how the Germans have made it work. And then we must make it work for us. I want you to go and catch me a Tiger.'

Doug was momentarily stunned. 'Are you sending me to Germany, sir?'

'No. You must go to, er...' Churchill stopped uncharacteristically in mid-flow. 'You'll know where you are when you get there. I'm going there myself shortly but I don't suppose anyone will allow me the time to capture a Tiger personally. So, Mr Lidderdale, you are my representative in this matter. I want you to bring me a Tiger tank. It matters not a jot to me how you go about your business. A butcher and bolt job will be perfectly acceptable.'

Churchill stopped and caught the gimlet eye of Major Desmond Morton. He looked at the tip of his cigar and then continued in a quieter manner. 'I want you to bring it here, Lidderdale. Park the bloody thing outside my front door. Do you understand?'

'Perfectly, sir.'

'I speak to you in confidence.'

'Absolutely, sir.'

Doug's mouth had suddenly gone dry. He swallowed and found he could say nothing more.

Churchill seemed unaware of his discomfiture. 'Have you ever been abroad before?'

'I spent some time in China.'

'All tropical diseases are much the same. Get a supply of Mepochrine tablets.'

'Mepochrine?'

'Anti-malarial stuff. And you can use our doctor here before you go. Save you a bit of time. Get yourself a typhus jab.'

'Yes, sir. When exactly am I supposed to be going, sir?'

Churchill had several large diaries near him. He consulted one and then turned back to Doug. 'A convoy is due to leave on the 22nd of this month. Be on it.'

NOEL BOTHAM AND BRUCE MONTAGUE

Doug nodded, too surprised to speak. He wondered if he should mention how inconvenient this was; that this assignment would leave him only a couple of days for a honeymoon. Then he decided that his misgivings might sound trivial in comparison to possibly winning the war.

Major Morton intervened. 'May I make a suggestion, sir?'

Churchill looked at him. 'Yes, Morton?'

'Wouldn't it be quicker to fly him out there, sir?'

Doug flashed the sort of look at Morton that should reasonably have been expected to reduce the man to ashes.

Churchill shook his head. 'Far too dangerous. Tank experts are an endangered species. We can't afford to risk losing one. Whereas I can fly anywhere – but then, Commanders-in-Chief are ten-a-penny.'

Churchill chuckled at his own flippancy, before continuing in a more serious tone. 'Even sailing is no picnic. Grand Admiral Doenitz's U-boats are hunting in packs of 30 or more. And Corporal Hitler has still got a worthwhile navy. No, there's a degree of safety in a convoy. You can zigzag and take evasive action as necessary.'

Churchill caught Doug's eyes in such an intense way that a casual observer might have thought the Prime Minister was trying to read his mind.

Finally, he said, 'You may not always be able to report to me. Your responsibility will be direct to the Cabinet. However, there will be someone there to look after you. Mr Morton will arrange things for you.'

Churchill stood up. The interview was over.

Morton and Lidderdale jumped to their feet. Doug replaced his cap and saluted smartly.

Churchill began to relight his cigar. As the men turned to leave, they heard him ask who was next.

One of the typists looked in her shorthand notebook. 'Mr Anthony Eden wants to see you, sir.'

The old man must have been dreadfully tired. Churchill had suffered a number of minor heart attacks which would not be revealed until much later. Now, he lived up to his reputation as a tenacious bulldog. 'Send him in!' he barked.

As Doug closed the door, he caught his last glimpse of Winston Churchill raising the snifter glass of brandy towards his lips.

A sombre man with the air of an undertaker was waiting in the corridor.

'Major Lidderdale, this is Sir Charles, the Prime Minister's physician,' said Morton.

'If you'd be kind enough to follow me, please?'

'I'll wait out here for you,' Morton told Doug.

Doug walked with the doctor into his small consulting room. Everything was laid out ready for him. He removed his coat and rolled up his right sleeve, for Sir Charles Wilson to take up the hypodermic and administer the typhus jab.

'This one's a bit nasty, but the next two, which you'll need at weekly intervals, won't be too bad. They'll also probably issue you with some Mepochrine anti-malarial tablets and don't forget to take them. It still kills more people in the world than all you chaps put together can manage.'

Doug rubbed a painful arm as he followed Major Morton to the spiral staircase. He stopped a moment to look at the poster that had been pasted beside the Marine standing guard at the foot of the steps. The sign read:

BE LIKE DAD
KEEP MUM

CHAPTER FIVE

8 JANUARY 1943 –1830 HOURS

The staff car was waiting in King Charles Street. A female driver was dressed in the uniform of the Women's Royal Naval Service – double-breasted jacket and skirt with a shirt and tie. She got out of the car and opened a back door.

Major Morton said, 'I won't be coming back with you. Leading Wren Driver Mackintosh will take care of you.'

Doug looked at the inscrutable older man and said, 'Thank you, sir.'

'What for?'

'You chose me for this mission, didn't you?'

'My dear fellow, I am merely Mr Churchill's adviser. But the main thing is that he likes you. What is more, he means what he says. And he means to receive what he asks for. So, good luck, Major.'

Doug held out his hand but Morton didn't shake it. Instead, he mumbled, 'In my estimation, you have just enough time to pack your things and start putting a team together.'

'What team?'

'You will need some men on whom you can rely. You're not going to steal a 60-ton tank by yourself.'

'Where are you sending me?'

'You will be made aware of your destination once the ship has set sail. But you need to select two or three key men before you leave. Who knows what you'll find at the other end.'

'You're right, of course. I must recruit a couple of aides. My mind hasn't quite come to terms with all this yet.'

'Well, you'd better concentrate, old boy.'

'I was thinking of my fiancée.'

'Ah! The future memsahib. You must, of course, call the little lady and put her in the picture. Using discretion, of course.'

'I'll do better than that. I'll drive down to Folkestone to see her.'

Morton tried to read the luminous dial of his watch and failed. 'Can't she come up to King's Lynn to see you? After all, you won't have much time before leaving for Liverpool.'

Doug looked up as if he had been shot. 'Liverpool!'

'That's where your ship is leaving from.'

'How far is it to Liverpool?'

'About 200 miles.'

'And the ship leaves when, exactly?'

'She sails on the 22nd. Leading Wren Driver Mackintosh will ensure you're packed away in time.'

Major Desmond Morton turned on his heel and disappeared into the darkness. Three years later, he would receive a knighthood.

Wren Driver Mackintosh parked the car in King's Lynn as the air raid sirens wound up and started blaring. Mackintosh and Doug hurried down to the recently built Anderson shelter.

Doug was fretting. 'I've got to get to a phone,' he said.

'You'll have to wait for the all clear,' the Wren reminded him.

When the sirens sounded again, Doug ran to the nearest phone box, inserted his penny, rang the number and pressed Button A.

'Hello, darling, I'm sorry it's so late,' said Doug breathlessly.

The woman's voice at the other end replied, 'This is the operator speaking for the South Coast District. All the lines are out of order at the moment.'

'This is a bit of an emergency,' said Doug. 'I'm an army officer speaking from a tank unit in King's Lynn.'

'I'm very sorry, dear. All the lines are down. Our engineers are doing the best they can. If you press Button B, you can get your money back, sir.'

'Thanks.' Doug reluctantly did as he had been told, retrieved the large brown penny and put it back into his pocket.

It took two days to restore the telephone connection. Finally, Doug got through and explained the situation to Kathleen as best he could. Knowing that the lines of military installations were prone to phone taps by foreign spies, Doug could not give all the details of his assignment.

'Tell me where they're sending you,' asked Kathleen.

'You know I can't say too much over the phone.'

'Well, when then?'

'In exactly two weeks time.'

Kathleen gasped. 'How long will you be gone for?'

'I don't know.'

'Why the haste?'

'It's a mission, darling ... I don't want to sound melodramatic, but it's a secret mission.'

'Douglas! You're not a spy. You're an engineer.'

'Yes, dear. Look, it's difficult to explain just now ... I was hoping you could drive up here before I have to go.'

'When?'

'Now. Can't you come up now?'

'I was saving all my petrol points for the honeymoon.'

'I'm so sorry. Beg, borrow or steal some more points from somebody. I'm in the old block here. They never got round to putting me into married quarters. Just do your best.'

There was a moment's silence and then the pips went.

Kathleen said, 'I love you, darling.'

'I love you like crazy, darling.'

'Bye now.'

'Do your best.'

With that, they were cut off.

Kathleen managed to scrounge sufficient petrol coupons to enable her to drive the 190 miles from Folkestone to King's Lynn. Pelting rain slowed her journey and she did not arrive until early the following evening.

Meanwhile, Doug began considering who from the 25th Tank Brigade might be recruited to his team of Tiger hunters. He had already made up his mind to take Corporal Bill Rider as his tank driver; his cheerful cockney disposition had gone a long way to lightening Doug's workload since he became attached to the 25th.

Doug sat behind his desk as Corporal Rider opened the door to let in a large man with immaculately polished boots and a handlebar moustache. The squat but immaculate Sergeant marched into the room and saluted according to the rulebook – longest way up, shortest way down.

'At ease, Quartermaster Sergeant.' Doug spoke the order with the hint of a grin.

'Sah!' The right hand came down like a jack-knife and disappeared behind his back as the feet separated by ten inches.

'Stand easy. This is an informal interview,' said Doug.

The Sergeant allowed himself to bend a little and took a step closer. 'If this has anything to do with the requisition of driving shaft belts, sir, I want to make it known that the matter is being rectified.'

'Relax, Sergeant, this has got nothing to do with the Quartermaster's Stores.'

'I am an Artificer Quartermaster Sergeant, sir.'

'And a darned good one.'

'My function is to repair and recover mechanical and electrical equipment.'

'That is precisely why you're here. Your reputation goes before you, Sergeant. I'm told you can wave a magic wand and produce impossible things at a moment's notice.'

Artificer Quartermaster Sergeant (AQMS) Sam Shaw seemed unsure as to how to take this unexpected compliment. 'With respect, sir, I am not a bleedin' fairy,' he said after a moment.

'Certainly not, Sergeant. My goodness, I see I'm going to have to choose my words carefully when I talk to you. I meant to imply that you can perform miracles when spare parts go missing.'

Mollified, Sam Shaw smiled. 'I do my best, sir.'

'I believe you are already acquainted with Corporal Rider?'

'Indeed, sir.'

'He is going to be my driver.'

'You couldn't do better, sir.'

Corporal Rider grinned broadly.

'I've been given a special assignment, and I need someone with your – how can I put it? – improvisational skills.'

'I can repair any tank you name, sir.'

'Good. What about a Tiger?'

Sergeant Shaw blinked. 'I don't do animals, sir. I am not a bleedin' vet.'

'The Tiger is the latest German tank.'

'I've heard of it, sir. Upgraded Panzer, ain't it?'

'Nothing like any tank you've ever heard of, Sergeant. This is completely new and has only come into service in the last few months. It's about 27 feet long and about 12 feet wide. Supposed to weigh something over 60 tons. Oh, yes – and one other thing – it can travel underwater.'

'If you will permit me to speak, sir?'

'Carry on, Sergeant.'

'Bloody 'ell, sir.'

'I think you speak for the three of us, Sergeant.'

'Where is this tank, sir? You want me to strip it down?'

'Eventually. We've got to capture one first.'

'Capture one? Where, sir?'

'How would you like to come on a hunt for one with us?'

'It sounds like an adventure, sir.'

'That is exactly it, Sergeant. Well, if you're willing, you'd better get your jabs and pack up your old kitbag.'

'What? Now, sir?'

'Yes.' Doug rubbed his upper arm. 'And I warn you, these bloody jabs start to hurt after a bit.'

Sergeant Sam Shaw seemed to lose some of his verve. 'Jabs for what, sir?'

'Tropical diseases. Oh, I say, didn't I tell you? We're leaving for a tropical destination which can't be identified.'

Sam Shaw seemed stunned. 'When do we leave?'

'Less that two weeks.'

Shaw's jaw dropped. It took a lot to prick his bubble, but this had done it. He gave his huge moustache a reassuring tug.

'There's one more thing,' said Doug.

Sam Shaw's demeanour diminished a little further.

'Sergeant, this is a top-secret operation. You will have to sign the Official Secrets Act.'

'If it's skulduggery you're after, I think you've chosen the right men for the job,' said the NCO with the beginnings of a satisfied smile.

Doug stood and put out his hand. Sergeant Shaw took it and numbly shook hands with his new boss.

'Welcome aboard, Sergeant Major.'

'Sah!' Shaw saluted automatically, spun on his heel and marched out in a daze.

At the door, he bumped into an attractive blonde in the act of shaking out her umbrella. He made his apology before heading off to seek out a medical orderly.

Kathleen ran into the room and into Doug's arms. 'Oh, darling,' she said, 'I thought I'd never get here. What on earth's going on?'

Doug turned to his Corporal and said, 'This is my fiancée, Corporal. Kathleen, Bill Rider.'

Bill said, 'I consider the Major is a very lucky man, ma'am.' He was a guileless charmer. 'Pleased to meet you, Mrs, um, I mean ... Miss.'

The Corporal retrieved his beret and walked to the door. 'Would you excuse me, sir? I just want to see a man about a dog.'

As soon as the door was closed, Kathleen said, 'What dog does he mean?'

'It's just an expression, dear. It was his tactful way of leaving us alone together.'

They kissed. It was a kiss they had both longed for ever since Doug had been so unceremoniously whisked away three days earlier. Only two people deeply and hungrily in love could have had such a kiss.

Afterwards, Kathleen felt her knees tremble a little. 'Douglas, dear,' she whispered, 'I think I'm going to have to sit down.'

'I've booked us into a hotel,' said Doug.

'Let's go, shall we? And you can tell me everything.'

'Is it all right if we take your car?'

'Sure. But I don't know how I'm going to get back to Folkestone. I can't get any more petrol for weeks.'

'Don't worry. I've got an Artificer who can fix that.'

'You've got a what?'

'I'll explain it all to you in bed.'

They went outside and ran through the rain to Kathleen's car. On the way, they passed Leading Wren Driver Mackintosh, umbrella up, hurrying towards the hastily erected Nissen huts that served as offices. The Wren saluted Doug who responded but neither exchanged words.

Kathleen shouted above the sound of rain, 'Who is she?'

Doug called back, 'My minder. Don't worry. She will never know where we're going to be hiding.'

An assumption Doug would discover to be entirely wrong.

CHAPTER SIX
19 JANUARY 1943

The Casablanca Conference was successfully kept secret until 24 January when it was announced at a press conference. Churchill wrote later, 'The press could hardly believe their eyes and then their ears when they heard the conference had been meeting for two weeks.'

The conference was a meeting between Churchill and Roosevelt and the two Free French leaders, General Charles de Gaulle and Henri Giraud, to decide where and when the Allied invasion of Europe would take place. Stalin had also been invited but could not attend.

After their meeting the leaders drove to Marrakech, where Churchill showed President Roosevelt the Atlas Mountains. When the President headed back to Washington, Winston stayed on and enjoyed his only opportunity during the war to do a spot of painting. He was a consummate artist.

Meanwhile, Douglas and Kathleen were furthering their wedding plans.

Major Douglas Lidderdale's hometown was Folkestone, a

coastal resort in Kent adjacent to the Cinque Ports, important naval strongholds since the Norman invasion. It also boasted 22 churches, but none of them would guarantee a time-slot for a wedding because all the ports in Southern England were subject to lightning bombing raids by the Luftwaffe at frequent and unpredictable intervals.

Consequently, on a bitterly cold Tuesday, 19 January 1943, it was in London, at Paddington Register Office that Aubrey Douglas Lidderdale married his sweetheart, Kathleen Crane.

As he placed the ring on the bride's finger, Doug caught sight of his mother, Betty, sitting on the front bench. A tear slowly rolled down her cheek, but she was smiling.

'I now pronounce you man and wife ... You may kiss the bride.'

Afterwards, the newlyweds shivered on the steps outside the register office while the photographer struggled with his bulky camera and flash bulbs.

Douglas Lidderdale stood six feet tall; a lithe and handsome soldier in his army major's dress uniform. Strictly self-disciplined both in body and mind, on this occasion, he was incapable of preventing his blue-grey eyes from straying in admiration to his bride, one of her arms clutching his, radiant in her wedding dress.

When Betty joined them for a group shot, she whispered to Doug, 'Your father would have been so proud.'

Doug gave his mother's hand an affectionate, reassuring squeeze. 'He is proud, Mother. His presence is always with me.'

It was dark by the time they arrived at the outskirts of Folkestone. The blackout restrictions meant Doug had to drive The Beast with masked headlights, which slowed him down. This did not sit easily with his natural instincts as a racing driver.

'I'm going to have to garage The Beast when I get back to the depot,' he said. 'I can't afford to run it now I'm a married man.'

Kathleen snuggled against his shoulder. 'You won't find me all that expensive,' she said warmly.

Doug braked and turned into the gravel driveway of a half-timbered hotel called The Tudor Tower.

'Are we having a drink before we go home?' asked Kathleen.

Doug opened the passenger door and helped her out before yanking their suitcase from the back seat. 'We're not going home. Not until tomorrow.'

They were shown up to a small en-suite room with a rickety floor and worm-holed beams across the ceiling. A four-poster bed with curtains dominated the room.

Later, as they prepared for bed, Kathleen sat on a stool in front of the dressing table, combing the confetti from her long blonde hair. Reflected in the mirror, she saw Doug standing behind her. He dropped his hands onto her shoulders, and then gently slid his hands down her neck to meet across her breasts.

'I love you,' he breathed.

She took his hands, raised them to her lips and gently kissed his fingers.

As she started to disrobe, she thought of what he had said to his mother earlier. 'Do you still miss your father?' she asked.

Doug frowned. 'Of course I do. I meant it when I said he is always with me.'

He told her the story of how his father – a dedicated doctor – had diagnosed liver cancer in himself a decade earlier. 'He announced the news to us that same evening over dinner. Mum took it on the chin at the time. She'd been a nurse – they had both trained at Guy's Hospital. In fact, I think that is where they met.'

'What about your brother and sister?'

39

'We were all stunned, of course. But then Dad said something that made the most profound impression on me. It's stayed with me to this very day.'

Kathleen paused and waited for him to continue.

Doug grinned. 'Just because I'm raking up old stories from my ancient past doesn't mean you can stop doing the striptease.'

Kathleen said, 'I may have been one of C.B. Cochrane's Young Ladies, but we never did anything crude.'

'There's nothing crude about getting undressed. We're a respectable married couple now ...'

Kathleen shimmered provocatively. Then, slowly, she removed her slip sensuously and let it fall to the floor. All that remained to cover her ripe, yearning body were the white stockings, the panties, the bra and the garland in her hair.

Doug approached and removed the garland.

Kathleen spoke softly. 'So what did your father say?'

'He said every day is so precious that we must live each hour as if it is our last.'

'Is that always possible?'

Doug took her in a close embrace and kissed her. Then, as he fumbled for the clasp of her brassiere, he whispered into her ear: 'I'll demonstrate.'

Twenty-four hours later, Kathleen looked anxiously at the petrol gauge of her Tin Lizzie as they drove out of the tank depot towards their King's Lynn hotel. They had only had one glorious night in The Tudor Tower love nest.

Doug broke the silence with a gentle jibe. 'You must have found it difficult to get the petrol for this old banger, right?'

'How dare you call my old banger an old banger! My Tin Lizzie gets me everywhere. Anyway, darling, if this is such an old banger, why aren't we in The Beast?'

'It positively eats fuel. I've had to mothball it for the duration.'

It was well past midnight and the honeymoon couple had gone some way towards making up for lost time. Lying in each other's arms with the lights out – for once they were only too willing to respect the blackout regulations – they whispered endearments that can only be passed between lovers.

And then there was a knock at the door.

'Don't answer it,' hissed Kathleen

'I don't intend to.'

Leading Wren Driver Mackintosh's voice rang out. 'Sorry to disturb you, Major, but something important has cropped up.'

'You don't have to answer,' whispered Kathleen. 'Pretend we're asleep.'

'Through that din?' said Doug, as the knocking redoubled in volume. 'Anyway, Churchill might not approve.'

'How will he know?'

'Well, the truth is, he gave me my orders.'

Kathleen sat up. 'Churchill? You mean you've actually met Winston Churchill?'

'Yes. He's the one that's sending me away.'

'Well, you'd better answer the door.'

'Are the blinds tight shut?'

'Yes.'

'Put on the bedside light.'

Kathleen cocooned herself in the bed-sheets as Doug struggled into his trousers and opened the door.

'Good evening, Wren Driver. Or is it morning?'

'It is morning, Major. Urgent news. Your ship is under orders to embark.'

'From Liverpool?'

'Yes, sir.'

'I was told I had 24 hours.'

'The time has been moved up slightly and you and the 25th Tank Brigade are leaving by train this morning.'

'I'll never get a train at this short notice.'

'The rest of your unit is already on its way, sir. My orders are to drive you to Euston. You will travel on a troop train from there.'

Doug gave a long sigh. 'How did you know where to find me?'

'That's my job, Major.'

'And the train goes direct to Liverpool?'

'It terminates at Liverpool Riverside. I think there will be an overnight regrouping at a transit camp outside Liverpool, and then your ship sails from the Prince's Landing Stage.'

'Are you coming with us?'

'Oh, no, sir. I'm Special Ops. I go back to my HQ.'

'Great Scot! I've left all my stuff back at the mess. I've only got a toothbrush with me.'

'I took the liberty of instructing your batman to pack for you, Major.'

Doug stared hopelessly at the Wren. Resistance was futile. 'Perhaps you would be gracious enough to give me five minutes to say goodbye to my wife?'

The uniformed NCO replaced her hat, gave a sympathetic smile and withdrew.

Doug returned to Kathleen. She was weeping unreservedly.

'I'm so sorry,' he whispered, lifting her chin up and kissing the tears from her cheeks.

'Oh, Douglas,' she sobbed. 'Blast this bloody war.'

'I love you.'

'How will I know if you are safe?'

'I'll get messages through somehow. Coded if necessary.'

They embraced. They could hear the engine of the car outside.

He whispered, 'I love you, Kate.'

She shivered. 'Bon voyage. Write to me every day.'

'I will.'

'Remember your father's words.'

'Hmn?'

'We must live each day as if each hour is our last.'

Those were the last words they would say to each other for nine long months.

CHAPTER SEVEN

22 JANUARY 1943

A howling wind drove the torrential rain across the open quays of Liverpool's South Docks to lash at the sides of the ancient military bus as it cleared the gates. The wind sent the bus reeling; its primitive suspension groaned as it struggled against the full onslaught of the winter gale driving in across the Irish Sea.

Doug and his men had spent the night with members of the 25th Tank Brigade, who would be sailing with them, in an anonymous military holding station on the edge of the city. The atrocious weather, combined with a strictly observed blackout on their earlier journey through the suburbs, meant they had seen little of the horrendous bomb damage Liverpool had suffered.

Now, peering through the streaming windows, which threw the outside world slightly out of focus, the men were staggered by the devastation they saw around them. Even the added filters of a leaden sky and early-morning January gloom couldn't hide the results of the Nazi Luftwaffe's assault on the city – a stark reminder of how Hitler's bombers now concentrated their raids on England's inner cities, waging war against innocent civilians.

Scores of warehouses and dozens of cargo and troop-carrying ships and barges had been gutted or demolished by more than 60,000 incendiary and high-explosive bombs dropped over just two December nights. Over 80,000 tons of goods had been destroyed; the whole port left in tatters.

Doug could see the flame-blackened superstructures of the sunken ships, pointing upwards like charred and gnarled limbs of lightning-ravaged giant oaks in the surrounding dock areas and in the Mersey. Some were ringed by huge cranes, some minus their top jibs and control cabins, while others leaned drunkenly on shattered supports or had been completely toppled. All were tinged reddish brown or black by the searing heat of incendiaries.

From behind Doug came the voice of AQMS Sam Shaw. 'Christ, sir, the bastards have been busy, if you'll excuse my language. Most of the poor bleeders in those boats and sheds must have been blown to bits or roasted to death when Jerry came calling. They wouldn't have had any warning at all. It must have been just like this in Pearl Harbor, sir.'

'You're right, Sergeant,' said Doug. 'Though I believe the Americans lost a damned sight more ships and sailors than these poor devils. But this is as bad as I've seen it, and it makes the success of our mission more vital than ever. Just imagine the carnage the Tigers could bring to our cities if they were given the opportunity. We must capture one of them and discover their Achilles' heel so our chaps can clobber them.'

'Amen to that, sir,' said the young Lieutenant sitting across the aisle from Doug. 'From what we've heard those Tigers are giving our tank lads a real bloody pasting. They can smash one of our Churchills from over a mile away and we can't even dent their armour from a quarter of that distance. It's a perfect killing machine, and that's going to become a big damper on morale. Whatever the Bosch has put together in that monster, we need to know it too – and quickly.'

Doug nodded in agreement, and shot a questioning look at the young officer who seemed so precociously bright and eager to get to grips with the enemy.

'Oh, I'm Lieutenant Whatley, by the way, with the 25th Tank Brigade workshop. Joined the outfit three weeks ago. Just in time to catch all the fun. Reg Whatley.'

Doug introduced himself and carefully examined the junior tank engineer officer. Reg had the chiselled looks of a matinee idol, and the lanky litheness of a long-distance runner. But he was totally unaware of his obvious charm and presented a diffident, almost self-effacing manner. Doug mused that Reg probably wouldn't let one down in a dangerous situation.

'I look forward to speaking to you again during our voyage, Lieutenant. Wherever we are going to.'

'We're being issued with tropical kit, sir.'

'True, but the tropics cover a wide area,' Doug pointed out.

He was cut off by the sudden opening of the door. An icy blast of wind carried rain viciously into the bus, producing a chorus of curses from the three dozen or so uniformed passengers.

The figure holding back the door was clad in a faded yellow sou'wester, a balaclava and knitted woollen gloves.

'Whatcha, whacks,' a rich Liverpool accent greeted them. 'Are you the tank blokes and the REME lot?'

The seaman's accent, with the added distortion of being muffled behind the balaclava, was barely decipherable to the southerners.

Doug understood the 'REME' part – Royal Electrical and Mechanical Engineers – and nodded.

'Is that our boat?' he said, gesturing towards a large two-funnelled vessel moored to the dockside 50 yards away.

'Ship, sir,' the strange figure corrected. 'And, aye, you'd better be looking sharpish 'cause the captain's keen to get the full morning tide and get out of the river before dark. Your baggage is being offloaded now and will be stowed aboard in ten minutes.'

Doug reached for his large holdall from the overhead luggage rack. As he discovered he was the ranking officer, he called down the bus to his team and the tank operatives who would be travelling with them to the tropics.

'Don't forget to grab all your hand luggage and kitbags, then get yourselves over to the ship directly in front of us at the double.'

He turned to Lieutenant Whatley. 'Come along, Reg, it's time for us to get wet.'

The bone-chilling, freezing rain almost took Doug's breath away. He tugged up his collar with his free hand and, head turned away from the direct force of the wind, rapidly strode towards the dockside, carefully navigating around some of the deeper puddles.

Behind them, he could hear one of the NCOs marshalling the men into a column to quick march them to the waiting vessel.

The ship which would be their home for the next couple of weeks had once been a liner but had been converted into an armed military transport vessel and was now painted from bow to stern in a drab dark-grey colour. Her name, the SS *Duchess of York*, was barely discernible on the soaring bows beside the anchor.

Corporal Rider shouted, 'She's named after the Queen, God bless her, so let's bleedin' pray she's just as tough as old Lizzy.'

This observation received a chorus of mainly four-letter assent from his fellow squaddies.

At the foot of the gangplank, their papers were cursorily inspected by a bedraggled rating who then directed the men up the walkway to a shell door in the ship's side, leading to a lower-deck vestibule.

There, a junior officer re-examined their documents and gave Major Lidderdale a smart salute.

'Welcome aboard, sir. You are on deck six and so is Lieutenant Whatley here. Almost neighbours. The rest of the men are berthed on lower deck two. No portholes I'm afraid, but there is an "other ranks" exercise area on the foc'sle deck. That's at the front of the main deck on level one.'

Sam Shaw rolled his eyes and prodded Bill Rider with his elbow. 'Did you understand that, William? The forecastle is at the front of the ship.'

'You know me, Sarge,' answered Rider with a grin. 'The only castle I know anything about is the Elephant and Castle, and it would make the officers blush if I was to tell you the kind of exercises they get up to down there.'

Doug laughed. 'One day you'll have to tell us more, Corporal. Meanwhile, go and find your berth and get something to eat.'

An hour later, having unpacked his kit, Doug joined Reg Whatley in the officers' mess – the former liner's first-class restaurant – for breakfast. As they had been advised on the dock, the ship's captain was eager to get away. An infantry company had followed them aboard and, as soon as the newcomers had been processed, the 20,000-ton troop vessel weighed anchor.

Even though they were still in the River Mersey, the strong wind buffeting the ship and the angry white-topped waves rolling in from the open sea were causing the old *Duchess* to pitch and roll. Many of the land-based passengers began losing their colour.

Lieutenant Whatley was one of them. He had refused food and opted for a mug of tea, but on the arrival of the Major's breakfast – an omelette made from dried egg, with fried tomatoes and toast – the younger man's face turned ashen and, clutching his hand across his mouth, he dashed from the mess.

Doug grinned at the Lieutenant's retreating back and silently thanked God he had been born with a strong stomach. He knew he would need it, as one of the ship's officers had told him they were currently experiencing some of the worst winter storms for years. It was going to be a pretty bumpy ride. High seas and gale-force winds would force the convoy, which they were due to join 200 miles north, in the Firth of Clyde, off Greenock, to spread out to avoid collisions, and this would make them easier prey for the so-called Wolfpacks of German U-boats awaiting them in the North Atlantic. In 1942 alone, over six million tons of shipping had been sunk by U-boats in what Churchill dubbed 'The Battle of the Atlantic', a battle that the Allies were currently losing, a fact of which Doug was depressingly aware.

'If the U-boats don't get us, there's always the bloody dive-bombers, flying out of France, who'll enjoy a pot shot at us,' an officer had warned nonchalantly.

Later, the same officer urged Doug not to be too downhearted. 'There'll be nearly 20 ships in the convoy and 17 naval escorts to shepherd us safely down to Gibraltar – including an aircraft carrier and three destroyers, so we'll have to be very bloody unlucky to get clobbered by the Krauts.'

The following morning, Doug staggered to the mess after a restless night's sleep during which the occasional crashing roll of the 600-foot ship had nearly tossed him to the floor. The mess was sparsely attended – even fewer non-naval officers had made it to breakfast now they were facing the full blast of the storm-force winds.

He was surprised to find Reg Whatley already seated and nursing a large mug of tea in both hands. The young Lieutenant still looked pale but managed to raise a grin when he spotted Doug.

'Morning, sir. I won't say "good" for obvious reasons. How did you sleep?'

'Badly.' Doug grinned in return.

His first impressions of this Tank Brigade Lieutenant were proving spot on: he had determination and guts when they were needed.

Doug ordered tea, toast and preserve, and glanced out of the port-side window. About a mile away, a large mass of land formed a smudgy, grey silhouette against the turbulent seas.

Reg followed Doug's stare. 'That's the Isle of Aran. We're just coming into the Firth of Forth estuary, apparently. One of the naval types told me we'll be heaving to in a few hours while the convoy and escorts gather round.'

Doug sipped his coffee. 'I suppose we still don't know where we're going. We are called Convoy KMF 8, if it's of any interest. The "K" stands for United Kingdom. The "M" stands for Mediterranean. The "F" stands for Fast.'

Reg was suitably impressed. 'Are you sure about that, sir?'

Doug laughed. 'No, I'm not. But it sounds right. Whatever they call it, they tell me it's going to be a rough ride for a few days. I don't think we'll be moving very fast.'

As they finished breakfast, Doug pointed to a naval vessel cruising a few hundred yards off their port side. 'That ship's flying a French flag. I thought they were on the German side.'

On the next table, a ship's officer laughed. 'No, some of them came over to us in 1940. You can see she's also flying the red cross of Lorraine. That shows she's part of the Free French Naval Forces. That's the *Savorgnan de Brazza*, a French sloop – a bit bigger than one of our destroyers. She actually sank one of her sister ships, which was loyal to the Vichy Government, to join us and fight the Nazis. She won't be coming with us all the way, though, just as far as the northern tip of Ireland. The British Navy only uses her for local escort duty.'

Doug frowned. 'Why only as an escort?'

'Don't know. I think she's earned our trust. But how much is another question. Personally I'm glad she's out there. Just in case there's a U-boat on the prowl.'

CHAPTER EIGHT

DECEMBER 1942/ JANUARY 1943

Fourteen hundred miles to the south, as the ship carrying Doug and his companions clawed its way into the Irish Sea through freezing rain and huge storm-lashed Atlantic breakers, another German killer was stalking its prey in the dusty wadis of Tunisia's desert scrubland. The infamous Tiger tank, the snarling embodiment of German military might, had been in North Africa for less than two months but already its fearful power had spread dread among the Allied tank crews – and for good reason.

Neither the American Sherman nor the British Churchill tanks could penetrate the Tiger's armour, even at close range, whereas the Tiger's powerful and deadly accurate 88mm Kwk 36 L/56 gun could destroy enemy tanks from a distance of almost two miles. Its superior armour and firepower was so devastatingly effective that Allied tanks were fearfully reluctant to engage it in open combat.

The psychological fear of Tigers had rapidly become so widespread among Allied tank crews it been given its own medical name: 'Tigerphobia'.

The first Tigers, in tank battalion group Kampfgruppe Lueder, under its commander Major Hans Georg Lueder, had wreaked havoc among the Allied forces in December.

On 1 December – just a week after being landed at Tunis off the transport vessel *Aspromante* – two of the initial three Tigers had destroyed nine American and two British tanks.

The following day, attacking the town of Tebourba, one Tiger, operating with five Panzer IIIs, knocked out six Allied tanks. On 3 and 4 December, joined by three other newly delivered Tigers, Major Lueder drove back the Allies and took Tebourba. Of the 182 tanks involved in the battle, the Allies lost 134, most falling victim to the seemingly invincible Tigers.

For the rest of December 1942 and most of January 1943, the Tigers lived up to their awesome reputation, with Lueder's battalion destroying 100 Allied tanks and 25 anti-tank guns in a single day. But, as would-be Tiger tamer Doug Lidderdale set out from Liverpool to confront his quarry, a group of his fellow Englishmen were succeeding, for the first time, in penetrating the armour of one of these 60-ton monsters and disabling it.

It happened in the hilly Tunisian hinterland on the road to a small isolated town called Robaa where the Allied forces in the shape of the Buffs (The Royal East Kent Regiment), part of the British 8th Armoured Brigade, and two troops of the 72nd Anti-Tank Regiment RA, were dug in to prevent the Axis forces pushing westwards.

The Axis attack was spearheaded by the German 501st Panzer Battalion with Kampfgruppe Lueder deploying to the south of the main road, and of the Kebir-See (Lake Kebir).

Deploying to the north was a second Kampfgruppe led by Hauptmann Pommèe, Commander of the II. / Pz GR. 69 with six groups of Tigers under Oberleutnant Löse, reinforced with two anti-aircraft and engineer companies. The two forces containing Tigers and Panzer IIIs and IVs were intended to protect a frontal assault by infantry – the Panzer Grenadiers Regiment – with Italian artillery positioned to give covering fire.

Most of the British troops were asleep in the slit trenches and it was before sunrise, at 0600 hours, on 23 January, when sentries reported hearing the distant clanking sounds of a large number of tanks on the move.

An hour later, silhouetted against the dawn sun to the east, the leading Tiger tanks and a group of Panzer IIIs trundled over the hilltop, engines roaring, their huge bulk setting the earth aquiver and causing sand to trickle into the trenches.

The British commander allowed them to approach within 500 yards of his infantry before ordering the six-pounders in the hills to their rear and the north to open fire. Within minutes, two of the Panzer IIIs were blazing, their screaming crews coming under fire from the Buffs as they tried to bail out of their stricken tanks.

The surviving tanks were returning fire, trying to home in on the tell-tale flashes in the British artillery lines, and behind them came a second wave of tanks and the German infantry.

Then the unthinkable happened.

Whether intentional, or by sheer luck, one of the six-pounder shells struck the last road wheel on the right side of the hull of the leading Tiger. The shell failed to penetrate but forced in the armour behind the wheel, rupturing a fuel line and causing the engine to stop. In the midst of battle, and at tremendous risk, the commander of a Panzer III manoeuvred his tank close to the back of the Tiger and the two crews managed to attach a tow bar to the stricken vehicle.

At a snail's pace and under constant enemy fire, the commander dragged the Tiger off the battlefield. But it was a wasted act of bravery: the Tiger filled with fuel vapour and suddenly, whether from an electrical short or a spark from moving metal parts, this ignited, causing an intense fire that burned until dusk.

A second Tiger, which suffered engine failure, received a staggering 24 direct hits from Allied tanks and artillery as it was being towed out by a Panzer III and, incredibly, it survived – a tribute to its remarkable armour plating. Only five of the rounds penetrated the hull and these caused negligible damage to the interior. The wounded Tiger was fully repaired after receiving a replacement turret and was back in action within the week.

Allied mechanics had hoped to salvage the burning Tiger but, during the night, it was moved out of sight by a pair of heavy tractors under cover of darkness and stripped of all its useable remaining parts, including the running gear and turret. Detonating a 50kg explosive charge under the hull destroyed the remains.

When the mechanics located it, they found, to their great disappointment, that there was nothing left for them to examine.

However, the battle on that day had proved one important thing, invaluable to Allied morale: the mighty Tiger was not quite the unstoppable man-eater they had come to believe.

The hellcat was vulnerable and the British dogs of war had picked up the scent.

CHAPTER NINE
24 JANUARY 1943

As Doug Lidderdale moved around the ship that morning, he was accompanied by the overriding smell of vomit. By this time, most of the military passengers aboard the *Duchess of York* had been struck down with seasickness. The worst cases were among the ordinary troops billeted in cramped conditions on the lower decks with little ventilation.

But this didn't prevent the men of the North Irish Horse, a tank regiment which had joined the ship a day before Doug had embarked, from staggering their way to deck in the early hours and serenading their homeland as the convoy steamed past the northern coast of Ireland.

The singing – a moving, farewell lament from a great fighting regiment on its way, once more, to war – continued until long after the Irish coast was hidden by the huge, white crested waves which wreaked havoc on their stomachs but could not quell the spirit of these proud Celtic soldiers.

For those who still retained the desire, or the ability, to eat, the food aboard ship, Doug discovered – with a certain amount of guilt – was far superior to that generally available in war-ravaged, ration-book-restricted Britain. Steak, bacon and eggs featured regularly on the officers' mess menu. The one jarring omission, for officers and men alike, was the total ban on alcohol. KMF 8 was an Allied convoy and, as Britain's chief allies were the Americans – whose navy did not permit alcohol on board – the Admiralty had reluctantly agreed that, in deference to the Yanks, everyone would observe prohibition while afloat.

There were strong rumours among the crew that troopers in the North Irish Horse had set up a still in the bowels of the ship, but not a drop of the precious nectar made its way to the officers' mess and Doug remained a reluctant teetotaller.

By the afternoon of Sunday, 31 January, the weather had begun to ease. Doug and Reg were able to stand in shirtsleeves and enjoy the warm sunshine on the promenade deck as their convoy reassembled into separate groups.

Only the day before, they had finally been given their destination port – Algiers – and issued with a special War Office booklet of advice for North Africa-bound troops, including graphic and dire warnings against unprotected sex with the local female population – both Arab and European.

Eight ships, including the *Duchess of York*, accompanied by nine of the Royal Naval escort vessels, headed east towards the Straits of Gibraltar and the Mediterranean. The rest of the convoy, with the remaining eight escorts and the only aircraft carrier, continued steaming south towards the Cape.

It was after midnight when they passed Gibraltar, which was bathed in bright moonlight. It seemed to Doug that half of the troops aboard – most of whom had got over their seasickness – had stayed up to view the legendary British stronghold, and they were treated to an unexpected light show when a plane engine was heard and a dozen or more searchlights came on and swept the sky to the northeast.

This dramatic performance drew loud cheers of encouragement from the men who clearly felt it also warranted a round of applause.

The following morning, Doug and Reg stood on the promenade deck of the *Duchess of York*, leaning on the rail in silent companionship to view their approach into the Bay of Algiers. It was the first time either man had visited the dark continent of Africa.

The warm caress of the Sirocco wind, which came from the Sahara desert, gently touched their faces as they stared southwest to the harbour walls fronting Algeria's capital city of Algiers, which loomed larger as they moved at a snail's pace towards the docks.

Along the coast, a ship that had beached after having its stern blown off by a torpedo was one of the few obvious reminders of the war. Half a dozen mosques were visible on the steeply sloping hillside that rose from the sea front, their slender minarets and great white domes gleaming in the intense crystal-clear sunlight.

The most modern part of the city was spread along the seashore with the ancient section sprawling skywards around the giant casbah – the city within a city – and its narrow alleys and sturdy walls, some dating back to the Phoenicians and the Romans.

The two men could already detect the African smells wafting out to sea in the warm Algerian air. In the distance were the figures of Arabs in long robes, many of the women in veils, mingling with the Vichy French – until recently their enemies – and groups of Allied servicemen, all going about their business along the harbour promenade. Scampering among them were hordes of Arab children, mostly naked or in rags, some wearing bizarre outfits made from cast-off or stolen army sacks.

With the smell came the flies and mosquitos – swarms of

them. An unpleasant foretaste of what, for most of the men, would become the least attractive part of life in North Africa.

Doug was surprised to find that the town had suffered little damage when it was invaded from the sea by American and British infantry in early November the previous year. The French had offered only token opposition and some had thrown down their arms and welcomed the Allies ashore. Within hours, it was all over and the Vichy General commanding the city had surrendered. There were some French troops who preferred imprisonment to switching sides, but the majority had accepted their altered role with a Gallic shrug of indifference, a reaction which had evoked little surprise among the majority of the British military. Since then, the only real damage had been done by German fighter-bombers delivering a physical manifestation of Hitler's contempt for his former allies.

By mid-afternoon, most of the *Duchess of York*'s passengers had disembarked, each man with full equipment and kitbag, to march in long columns through the town's cobbled streets and up to 15 miles along dusty country roads to their respective transit camps.

Reg Whatley and his unit of the 25th Tank Brigade were scheduled to travel overland to a harbour between the Tunisian cities of Le Kef and Teboursouk. The harbour would soon be a gathering place for tanks, but the men travelled ahead of their fighting vehicles, which would be arriving on a later convoy. Reg and Doug, whose REME unit was destined to be in the same area, had agreed to meet there. The two men had formed a strong bond during the previous ten days and it was with reluctance that they had shaken hands on the dockside before parting. Unbeknown to the young Lieutenant, Doug had already earmarked Reg as a potential Tiger hunter.

'I know we're both here to help win a war, Reg,' said Doug,

'but I've been given another task to perform in Tunisia, and I think you are just the sort of chap who can help me. There's a risk involved, but if you're interested I'd like to talk to you more about it when we meet up in Tunisia.

Reg's eyes lit up. 'If it involves giving Jerry a damned good hiding, then you can count me in, sir. I'll look forward to seeing you there.'

Doug grinned. 'Oh, this will hit him where it hurts, you can count on that, old boy. Meanwhile, happy marching. I can't say I envy you the next few hours in this heat. I'm told that I've been lucky enough to get a billet in a hotel here in the city for a couple of nights before sailing on to Bone. And in a bed that doesn't constantly bloody rock, thank God.'

'Lucky blighter!' The Lieutenant laughed, holding out his hand. 'We're doing it the hard way. Slogging right over the Atlas Mountains. Our tanks will follow on, hopefully, by transporter or rail. But high road or low road we'll both be in the thick of it soon enough. Not much doubt about that.'

Doug could see Reg was supremely happy at the prospect of squaring up to the enemy, and this confirmed again that Whatley would be an excellent addition to his team.

As Reg strode off to join his unit, Doug was joined by Sergeant Shaw, Corporal Rider and the rest of his REME contingent, marching two abreast. A pair of junior officers followed them, out of step.

Shaw, who was turned out in immaculately pressed battledress – for the powers that be had still not deigned to issue the troops with khaki drill – brought the squad to a halt, and snapped off a perfect salute.

Doug acknowledged it with a salute of his own.

'Sah. Permission to proceed to our temporary quarters, which I'm told are about a mile along the front?'

Doug nodded, as he casually acknowledged the salutes of his officers. 'You have the orders of the day, Sergeant?'

'Yes, sah. We are to be back here, on the docks, at 0900 hours the day after tomorrow – Wednesday, sah – to board a destroyer for the next leg to Bone, about 300 miles along the coast.'

'Right. I'll probably drop in on you tomorrow, Sergeant. Until then try to keep the men out of trouble. And that goes for you chaps too.' He turned to the two Lieutenants. 'I imagine drink is pretty cheap around these parts and there are plenty of other temptations to distract you – as you've no doubt spotted already.'

In the short time the men had been standing, a small group of dark-skinned teenage girls, some as young as 14 or 15, had gathered at the roadside, making obscene motions with their hands and mouths and beckoning the men towards them. To leave the men in no doubt as to their intentions, a couple of them, legs splayed, lifted their ragged skirts to show their naked lower bodies.

A few of the men laughed or called lewd suggestions but were quickly silenced by a barked command from Shaw.

'You can see the danger, Sergeant. But try to make them keep their trousers buttoned. I was told by officers on the *Duchess*, who know this place, of a little ditty invented by the Desert Rats, which they should all take to heart: "Pox does more than Rommel can, to Bugger Monty's battle plan." Let's try not to do Rommel too many favours, if you please.'

'Yes, sah. I'll take care of them as if they was my own, sah. You can count on it.'

'Right, Sergeant. Carry on and I'll be in touch.'

After a fresh round of salutes, the men moved off, followed by the ever-growing, raggle-taggle group of teenage prostitutes.

Doug looked around and spotted a battered Jeep with a RASC

private standing beside it. He was holding a cardboard sheet with 'Maj. Lidderdale' scrawled across it in red.

CHAPTER TEN

1 FEBRUARY 1943

Doug crossed the road and identified himself, and was soon being driven along the crowded boulevard of Algiers harbour at almost suicidal speed by his cockney driver, who looked barely old enough to have passed a driving test.

When Doug mentioned this, the private laughed out loud. 'Course I ain't! They stopped the bleedin' tests because of the war, didn't they?'

Doug clung to the open sides as they careered around the Arab food stalls and crowded pavement cafes. The atmosphere was abuzz with hagglers screeching at besieged vendors and cosmopolitan café customers sipping drinks. Doug marvelled at the ancient trams, which clanged and clattered along the main roads; people were hanging from the sides and on the roofs, and some brave souls were even perched on the front bumpers.

The driver was still laughing when the Jeep screeched to a halt outside the Aletti Hotel, a large white, rather grand building which in the past had been home to visiting heads of state. It was among the most chic of Algiers' addresses.

Doug stepped from the Jeep and was surrounded immediately

by half a dozen bedraggled Arab children, hands outstretched, offering shoeshines, cigarettes and a host of other products and services. They were shooed away by a burly uniformed doorman wearing a tall fez with a tassel. He seized Doug's luggage and, with a series of bows and grunts, ushered him into the reception area.

Twenty minutes later, Doug was installed in an enormous room with a magnificent view across the bay and anticipating a nourishing drink in the bar. He had been proudly assured by the receptionist that this opened between five and seven thirty pip emma, and was one of the few bars serving alcohol in Algiers.

He was therefore disappointed to find that the bar, though impressively ornate, had little to offer by way of refreshment. There were only a handful of other men at the bar, all in officer uniforms, but the fat, balding, middle-aged French barman, who sported a huge handlebar moustache, took several minutes to amble in Doug's direction.

'Beer?' enquired Doug.

'We 'ave beer, monsieur. No spirits and no aperitifs. We 'ave lemonade, ginger ale, a local red wine and water. Which would monsieur like?'

'A beer.'

One of the drinkers at the bar loudly cleared his throat. Doug glanced along the bar and saw that all the men appeared to be drinking red wine, though the shade of burgundy changed from glass to glass. A Major of about Doug's age saw him staring and called down the bar.

'The beer's so watered down it makes gnat's piss taste strong. You're better off with this rotgut. But take my advice, old boy, and add a little lemonade. It just about makes the stuff drinkable.'

Doug waved his thanks and decided to heed the man's advice. When he turned, he found the barman was already pouring wine from an unlabelled bottle into a tall glass. Then he produced a half-empty bottle of lemonade, with its screw-top missing, and slowly added the contents, which were devoid of even a single bubble, to Doug's glass, while giving him an enquiring glance.

When he had added about half the amount of lemonade as there was wine, Doug raised his hand. 'Let's try it there for starters,' he said.

'That will be one shilling,' said the barman, replacing the bottle under the bar.

'Better get some ice, old boy,' called the Major. 'Makes it slightly more drinkable – but only just.'

Doug grinned another thank you to his fellow officer and, while the barman splashed two cubes of ice into his drink, he sorted out a shilling from the change in his pocket and placed it on the bar. If this was anything to go by, the drinks in Algiers were not exactly cheap.

The barman regarded the coin and raised his deadpan face to look Doug in the eye. 'Is monsieur not 'appy with the service?'

'Not at all. It was fine.'

The waiter switched rheumy eyes down to the shilling and back up to Doug – and waited.

Doug finally twigged. The cheeky blighter was asking for a tip. He searched his trouser pocket again and picked out two pennies, which he added to the shilling.

The barman scooped the three coins into his hand and, with a muttered 'Merci, monsieur', he shuffled over to the till.

Doug sipped his drink and grimaced.

'You say the beer's worse than this?' he called to the Major. 'Then it must be pure sewer water.'

The Major moved along the bar and thrust out his hand. 'Ron Parker.'

Doug shook his hand firmly. 'Doug Lidderdale, REME. We just got in.'

'Welcome to hell,' said Parker cheerfully. 'This town shuts its doors at sunset and dies. Thing is to grab your bully beef in whichever dim little restaurant you can find and then head for bed. There are a few black-market restaurants about – but they also charge black-market prices. After that, most of us go to bed. And most go alone. The girls on the streets are young and willing, but also filthy, and capable of passing on a dose of the pox that half a ton of mercury and sulphur drugs wouldn't shift. We have the new penicillin but Churchill said to save that for the badly wounded. His personal instructions are that any man who catches the pox is to be shipped straight to the front line. That's his idea of effective treatment.'

Doug raised his eyebrows and listened as Parker continued.

'Trouble is most of the troops, ours and the yanks, are still virgins, and with fifty fags a week as part of their pay and sex costing five fags a go, the temptation for most of them is overwhelming. French letters are free but half the blokes don't even know what they're used for.'

Doug chuckled and he appraised the other customers for a moment. The Major was eager to impart his local knowledge. 'The women who work this bar, who are mainly French, are probably clean but they charge a week's wage for a quick tumble.'

Both officers risked another sip of whatever they were drinking before the Major continued. 'The only other option is to visit one of the dirty little bars in one of the dirty little alleyways, and there, my friend, you'll most likely be relieved of your wallet by one of the wily Arab natives who can't wait to see the back of us. There's a powerful smell of decay in Algiers. Not just physical but cultural as well. An abode of the damned; of tattered scarecrows with the morals of an alliance of psychopaths. Most would sell you their sisters – or mothers – for a few cigarettes.'

'You paint a pretty bleak picture, that's for sure,' said Doug. 'I'd always imagined it to be a romantic spot.'

'As romantic as a camel's backside, old boy. They call this the white city, but it's got a very black heart. The Arabs think nothing of digging up our dead and stripping them of their clothes and watches and the like. Even grab their false teeth.'

'Dear God.'

'Some of our chaps have been known to set the odd booby trap on graves and when the Burke and Hare Arab tomb robbers get to work – boom! – they quickly find themselves shaking hands with Allah a lot sooner than they expected.'

Parker drained his glass. 'The PM's son, Captain Randolph Churchill, was the first Allied soldier to reach this bar when they took the town three months ago. I hope he got a better bloody drink than they're serving us. There are a couple of posher places, but not much. What are your plans this evening? There's still an hour of daylight left.'

Doug looked at his watch. 'I must report to the local military headquarters in the Place du Gouvernement. After that, I'm free.'

'Couldn't be better,' said Parker. 'One of the few decent bars in town is only a quarter of a mile from here, right on that square, near the Grand Hotel du Louvre. It's called the Maison Dorée Café Restaurant. If you like, I'll stroll along with you and perhaps I can introduce you to one or two people.'

An hour later, after a brief stop at Allied headquarters – a series of huts on the pavement outside a mosque, where Doug had confirmed his onward travel orders to Tunisia – they were in the crowded and noisy Maison Dorée restaurant.

Parker had found them seats at a table occupied by a young RAF officer of his acquaintance – Joseph Berry, who hailed from County Durham. He was a tall, lanky, raw-boned man with a mop of dark sweptback hair, and reminded Doug of a young Leslie Howard. He was 22, but his experience of war as a fighter pilot, and the resultant strain in his eyes, had made him look older.

Parker introduced him as Joe. He was with 153 Squadron, based at Maison Blanche airfield, five miles along the coast from the city, and responsible for the night-time air defence of Algiers. They were equipped with Beaufighter aircraft, which at the time was one of the most heavily armed fighters in the world: four 20mm canons mounted in the lower fuselage and six 7mm Browning machine guns in the wings.

'Is it always this busy on a Monday?' asked Doug.

Joe Berry laughed. 'Is it Monday? Every bloody day's the same to us. Sometimes we have to scramble twice or three times in a night. Messerschmitts and Stukas, but Jerry rarely gets as far as the city; he's usually content to strafe the harbour and any ships that might be around.

'Once we show up, they scarper pretty quickly.'

Parker grinned. 'I suspect your mob has the same effect on the ladies.' Then, turning to Doug: 'It's about this full in here every night. The wine they serve isn't half bad. And they do it at two bob a bottle. Let me get the first one.'

He beckoned over a waiter and gave his order, asking for three glasses.

Doug looked around and saw that about three-quarters of the customers were in uniform, many of them Americans, including a trio of women who were also in American army uniform.

Joe followed his eyes and chuckled. 'Not much chance in that direction, old chap. Those ladies are probably the cleanest of their sex in North Africa but they tend to want to date only their homegrown fellas. Non-Yanks could buy them drinks all night and still not get to first base.

'Doesn't matter to me. I'm not interested. Been married less than a year and my Joyce is well worth waiting for. No harm in trying your luck, though.'

Doug grinned back. 'Not me. I only got hitched two weeks ago to the most beautiful girl in the world. I should really still be on honeymoon if it wasn't for this bloody war.'

The waiter arrived with their order and poured the wine.

Doug raised his glass. 'I propose a toast. To Joyce and Kate.' He turned to Parker. 'What about you, old boy. Is there a Mrs Parker back at home?'

The Major looked shocked. 'Good God, no. At least not any more, thank the Lord. But I'll happily drink to you two newlyweds and your brides.'

The three men clinked glasses and drank.

'Have you been here long, Ron, or are you just passing through?' asked Doug.

Parker lit a cigarette, inhaled deeply and grimaced. 'I'm rather stuck here, I'm afraid. Attached to Harold Macmillan's staff. He's Churchill's protégé – the resident British Minister and the old man's go-between with General Eisenhower.

'It's a political cesspit here, full of intrigue and suspicion. The French are now supposed to be on our side but Algiers is full of Vichy collaborationists who are still in positions of power. It's a complete nightmare. The Nazi race laws are still in force, would you believe, and prisons are still operating, concentration camp style, full of anti-fascists, Jews and gypsies, not to mention friends of the Allies and France.

'They'll get round to sorting it out sometime, but not until we've trashed Rommel and pushed Jerry out of North Africa. Until then, the powers that be are not interested.

'At least you two are getting a chance to do something positive on that score. Tanks and fighters, what? Let's bloody drink to those, and damnation to the Hun.'

The men raised their glasses again, took a deep breath and drained them in unison.

CHAPTER ELEVEN
FEBRUARY 1943

It was raining solidly when Doug and his men arrived in the eastern Algerian port of Bone, and there were reports that nearby Tunisia had suffered snow and bitterly cold winds, leaving the north of the country a muddy quagmire.

This was a land cursed with some of the harshest climatic and geographical challenges on earth. Temperatures could swing from a searing 120 degrees in the day to freezing at night. Sudden violent rainstorms could flood dried-up riverbeds – wadis – in minutes, washing away men and equipment – even 30-ton tanks. Blinding sand or dust storms known as Simoon could spring up just as quickly and the constant threat of poisonous snakes, spiders and scorpions and the world's most deadly killers – the mosquitos – were an ever-present threat to the unwary. Then there were the flies – a ceaseless source of irritation – and, through some nameless, faceless supply clerk's error, a constant lack of toilet paper.

Dodging bullets, Doug was about to learn, was one of the least of a soldier's problems in war-torn Tunisia.

After the comparative luxury of the *Duchess of York*, their passage from Algiers on a British Hunt Class destroyer had seemed cramped, but had passed off without incident. Except, Doug noted in his diary, for an aborted daylight attack from a couple of German Messerschmitt fighters which peeled off and dashed for home when a patrol of British Spitfires arrived, wing guns spitting streams of lead.

They caught only a fleeting glimpse of Bone, as their transport – a convoy of three tonners and staff cars – was already marshalled at the dockside to carry them on the final 150 miles to the 25th's Tunisian headquarters northeast of Le Kef, where Doug's 104 Army Tank workshop would be based.

Doug did notice that there was a good deal more bomb damage in Bone than in Algiers. The buildings that had escaped unscathed looked uncared for and unwashed. The mix of Allied military, French civilians and local Arabs seemed much the same, with, of course, the inevitable hordes of dirty half-naked children among the crowds.

He had left behind a Lieutenant and two men to supervise the unloading of spare parts, tank recovery vehicles and mechanical equipment from a following convoy, and didn't envy in the slightest their stay in this depressing city with its pungent smells of rotting waste and decay.

Their journey revealed some spectacular scenery. Easy on the eye, thought Doug, but an absolute bastard to fight in.

In the north, the mountains ran in jagged lines with deep, steep-sided valleys between them, while further south these opened out into lush, fertile plains with wheat fields, vineyards, olive groves and orchards of Tunisia's sweet tangerines, interspersed with sugar loaf hills and occasional mountainous spurs.

The hills and mountainsides were covered with dense, tall scrub and oak cork forests. There were few decent roads and, despite the efforts of gangs of Sappers, who were struggling in appalling conditions to widen and strengthen them, constant heavy military traffic and the recent rains had turned most highways into rivers of mud.

Accidents, and subsequent delays, were frequent, and all men were tired and irritable when they finally arrived at their destination near the ancient monastery of Thibar on the Allied 1st Army's section of the front facing Rommel's Afrikakorps.

After delegating AQMS Shaw and his junior officers to oversee that the men were fed and billeted, Doug grabbed a welcome mug of tea in the mess before reporting his arrival to Brigade headquarters. It was after midnight before he dropped on to his bed, fully clothed, and fell into a deep sleep.

It was still raining when he woke but his mood lightened over breakfast in the mess tent when an unexpected visitor joined him at his table.

'Good morning, sir. Welcome to Mudsville. It's a bloody quagmire out there.'

Doug looked up and jumped to his feet, hand outstretched. 'Reg, how nice to see you. When did you get here?'

Lieutenant Whatley pumped his hand and grinned. 'We slid in early yesterday, sir. The rainy season normally eases off by now, or so we're assured by the locals, but it's still bucketing down out there. How we are expected to operate tanks in these conditions, God only knows.'

'I expect it will be a week or so before all the equipment catches up. Then it's up to the powers that be. How did you know I was here?'

'One of our chaps up at Brigade headquarters mentioned you reported in last night. Not much else to do at the moment, so I thought I'd come and say hello.'

'If you hadn't, then I would have come looking for you,' said

Doug. 'You remember I talked to you about a special mission I've been entrusted with?'

Reg nodded vigorously. 'That's one of the reasons I'm here.'

'Well, I'd be very pleased if you'd agree to help me. It's not one for the faint-hearted, but from what I've seen you don't belong to that lot. It would be wrong though to pretend there isn't a risk involved. To be honest, it's likely to be bloody dangerous. We could easily get ourselves killed.'

'Sir, I knew what the risk was before I signed up. That I'd have to put my life on the line if necessary. So tell me what it is you want me to do. I'll take my chances as long as I know it's going to hit Jerry where it hurts.'

'No doubt about that, Reg. What I want you to do is to help me catch a Tiger. Not kill it or harm it, but bring it home alive – back to England.'

Reg Whatley sat back, a broad smile on his face. 'I do presume you're talking about a Tiger tank, Major. India's a bit of a hop from here if you're after the real thing.'

'Oh yes, Reg, it's the Tiger tank we're after. The Prime Minister wants us to bag one intact, then take it apart and use its secrets to improve our own machines, and hurt them back.'

'You mean the PM's behind this? Bloody hell! Old Winnie doesn't do things by halves, does he? Well, you can count me in, sir. Just one thing. How do we plan to go about catching a 60-ton tank without getting ourselves blown to Kingdom Come?'

Doug felt his smile widen. 'I haven't quite worked that one out yet. But I've no doubt at all that an opportunity will present itself. If we hang about the battlefields when the Tigers come out to hunt, and we are ready to take advantage of any hiccup on their part, I'm sure we'll get our chance.'

'Who else is on your team, sir?'

74

'So far, only Sergeant Shaw and Corporal Rider. Both good chaps. I'll square things with the Brigade commander at headquarters and then it's all a matter of waiting until we hear our Tigers are out hunting. That's when *we* become the hunters and *they* become the prey – hopefully. Meanwhile, just carry on with your normal duties. I don't think it will be long. Once all our tanks and equipment are here, HQ is going to want to get us involved. Rommel is being pushed out of Libya right now, and with Monty hot on his heels he's going to be looking for a way out.

'We're smaller than the Eighth Army and a lot of our chaps – particularly the Americans – are completely new to this game. I expect that very soon Rommel will come banging on our door and when he does it will be his Tigers that do most of the knocking.'

CHAPTER TWELVE
14–22 FEBRUARY 1943

Doug's prediction concerning the untested Americans came all too tragically true just one week later, in a Valentine's Day massacre more brutal than anything dreamed up by Al Capone and his fellow gangsters.

The surprise attack on the US communications and supply centre at Sidi Bou Zid came at dawn and under cover of a huge sandstorm. It was brilliantly coordinated. Four German battle groups, led by a large force of Tigers and under the personal command of Field Marshal Erwin Rommel, moved through the outlying Fa'd and Maizila passes with more than 140 German tanks from the 10th and 21st Panzer Divisions and completely routed two US infantry battalions and part of the 1st Armored Division.

The American Sherman tanks were no match for the deadly Tigers. The Americans were split into small groups and then systematically exterminated. A hundred US tanks and their crews were obliterated, along with dozens of half-tracks and big guns.

When the 168th Infantry's 3rd Battalion leader, Colonel Tom Drake, radioed HQ for permission to retreat, Major General

Lloyd, the US area commander, totally misjudged the situation. From his ultra-safe headquarters tunnelled deep into the nearby mountains, Fredendall ordered Drake to hold his position until reinforcements arrived. The promised reinforcements never arrived and over 500 of Drake's men died needlessly.

In all, during those first three days of fighting, 4,000 Allied troops were killed or declared missing in action.

By 19 February, Rommel had advanced 40 miles, his war-toughened troops having outmatched and outmanoeuvred the Americans who, though keen to fight, simply lacked their enemy's experience. It was a fight between ruthless professionals and poorly led amateurs, and it resulted in a bloodbath.

Time and again, the Sherman tank commanders would bravely rush to challenge the Panzer IIIs and IVs, only to have the Germans split left and right when confronted, to reveal the ferocious Tigers hidden at their centre. Totally outgunned, the American tanks would be reduced to scrap within minutes.

The psychological damage done to the Allies was immense. The news that Rommel had been given total command of Axis forces in Tunisia only served to raise anxiety levels even further.

The Desert Fox, the nickname given to Rommel by the British 8th Army, was set on taking Tebessa across the border in Algeria, which was strategically vital to Eisenhower's plans for a spring offensive against him. Fortunately for the Allies, the Commando Supremo in Rome, and the German High Command in Berlin, lacked Rommel's tactical wizardry and ordered him to strike against Thala and Le Kef to the north instead.

Rommel railed that it was an appalling and unbelievable piece of short-sightedness; that the long northward march would expose his flanks, leaving them open to attack. But he was forced, with disastrous consequences, to obey orders and split his forces. This turned out to be fatal.

Still cursing the stupidity of his stay-at-home superiors, Rommel reluctantly chose to make his main early thrust through the Sbiba Gap and on to Thala. The second thrust would penetrate the Kasserine Pass making a feint against Tebessa.

His strike through the Kasserine began late that afternoon, advancing through drizzle and mist and with heavy rain clouds obscuring the mountains on either side of the smooth but muddy uphill floor of the pass.

Both prongs of his twin assault met massive resistance from the Allies and, with limited visibility, there was ferocious hand-to-hand combat. Rain and low cloud also prevented aircraft of either side from playing an effective role, but continuous and deadly artillery fire sent the casualty rate soaring.

By noon on 20 February, the Americans and French, at a huge cost in lives, were beginning to falter, although still managing to inflict serious punishment on the Axis infantry and Panzers.

Finally, the French ran out of ammunition and were forced to retreat, leaving the Americans alone to prevent a German breakthrough. Their orders were as brief as they were simple: 'Fight to the last man standing. Die if you must but do not retreat.'

And that is what most of them did. Outmatched and outnumbered, they fought with incredible courage until they almost ceased to exist as a unit.

That afternoon, at 1630 hours, Rommel's men, led by the 8th Panzer Regiment, overcame the last remnants of the US infantry and surged from the Kasserine into the rolling countryside beyond. A squadron of Sherman tanks, whose commanders had chosen bravely, but unwisely, not to flee the battlefield, was trapped against a hillside north of the mouth to the pass.

From point-blank range, the Tigers and Panzers pounded the Shermans into a holocaust of blazing, tortured metal. Barely a single member of their crews managed to escape the carnage, and many were burned alive.

The following day, 21 February, Rommel ordered the original attack on the Sbiba Gap, though stalled by heavy resistance, to continue pressing forward. He hoped to keep the British defenders occupied and away from his offensive strike to the northwest.

Then he further divided his remaining force, sending the 10th Panzers north towards Thala and launching his Afrikakorps at the US defenders of Djebl Hamra, a large hill to the south which controlled the Tebessa road.

For more than four hours, the two battlefronts edged back and forth and some of the most bloodthirsty and merciless close-quarter combat of the whole North Africa campaign ensued. The infantry fixed on bayonets to take on the enemy face-to-face. Tanks blasted each other at point-blank range, and all around men were dying in their hundreds.

That evening, through sheer exhaustion, both sides ordered a stop to the fighting. It seemed inevitable to everyone that on the next day Rommel's forces must continue their advance. The Allies simply lacked the manpower to stop him. Then, a miracle occurred, in the unlikely shape of Brigadier General S. Leroy Irwin.

Irwin had been ordered by Eisenhower to rush his 9th Infantry Division's artillery to his beleaguered American countrymen's aid, and had travelled 800 miles non-stop from Morocco. Through driving rain and along virtually non-existent roads, the division brought 48 big guns – 105mm and 155mm howitzers and 75mm field guns.

Added to the 36 British artillery pieces, the fresh armament was enough to halt Rommel's dawn attack in its tracks on 22 February. The ferocious barrage knocked out his tanks, guns and trucks and decimated his infantry.

In the Sbiba Gap, Rommel's 21st Panzer Division had failed to dent the British and American defences and was also in trouble. Unaware that the Allies lacked any immediate reinforcements and were on the point of capitulating, and at the same time believing his offensive had lost its momentum, Rommel took the only choice he felt was open to him. He gave the order to withdraw.

A younger Rommel, the brilliant architect of a hundred successful battles, might have pressed on regardless and risked all for a decisive victory. But by now the General was a sick and tired man. With ammunition, fuel and food stocks dwindling and the High Command ignoring his pleas for fresh supplies, he no longer felt he had the backup to sustain an aggressive campaign. His biggest frustration was that, had the idiots in Rome and Berlin not forced him to divide his army, he could have taken Tebessa in just two more days, and dealt the Allies a devastating blow. All that was now mere fantasy.

On the morning of 23 February, he decided to abandon the entire offensive and sent word to the 10th Panzers and the Afrikakorps to pull back through the Kasserine Pass.

Even luck seemed to have turned against him. As Rommel's army retreated, the rain clouds that had obscured the region for four days suddenly vanished. Within an hour, more than a hundred Allied aircraft – including B-17 bombers, P-38s and P-39s and, from the RAF, Hurricanes and Spitfires – were bombing and strafing his retreating columns in a non-stop barrage, while smaller reconnaissance planes guided the Allied artillery fire on to his tanks and men.

Rommel had been stopped, but at a high price. The battle of Kasserine Pass added another thousand to the Allied death toll and caused a total of over 6,500 casualties.

It had one positive effect, however. The posturing, hard-drinking and cowardly commander of the US II Corps in Tunisia, Major General Lloyd Fredendall, whose lack of guidance, compassion and respect for his men had resulted in the unnecessary loss of hundreds of young lives, was axed. He was replaced by a genuine five-star hero – General George Patton – who would quickly turn his troops into a solid and effective fighting force.

At the same time, Britain's General Sir Harold Alexander – Churchill's favourite Allied commander, was given overall control of the two armies in Tunisia, the 1st and the 8th.

Now the Allies had two inspiring leaders, whom their men respected and trusted, and whom the enemy feared.

CHAPTER THIRTEEN
FEBRUARY– MARCH 1943

Doug's debut battlefield experience and his first chance to see his quarry in action came towards the end of February. The exposure to the savage reality of war and the Tiger's lethal power left him shocked and intimidated, but at the same time strangely exhilarated.

His initial reaction was that Winston Churchill had seriously overestimated his ability to deliver the goods. It would be impossible, he concluded, for a small group of men to capture one of these leviathans of the battlefield alone.

'The only way we can do it,' he told Bill Rider, who was at his side, driving one of the 104's six-wheeler Scammell heavy tank retrievers, 'is to let our artillery and tank chaps do the hard work for us.'

'Only way to do what, sir?' queried Rider, who was concentrating on negotiating their way through a mud-drenched, pitted slope in pouring rain and under enemy mortar fire.

'Catch our Tiger, Corporal, that's what. We must keep our eyes skinned – spot when one's in trouble and then pounce. No other way.'

Rider looked relieved. 'I was hoping the plan might be something like that, sir. I must say, those Tigers put the fear of God into me. That they do.'

Doug Lidderdale smiled. 'Don't worry, Rider, they do the same to me.'

During the following month, all units of the 25th Tank Brigade were involved in fierce running battles along a 20-mile front, as the Germans tested their strength from El Aroussa in the South to Beja in the North.

This was the first experience of war for the majority of troops serving in the tank regiments, including Doug. It was far from an easy baptism. They were up against toughened, highly experienced Axis veterans who asked for, and gave, no quarter. Any romantic notions that tank battles would be repeats of the fabled air-ace dogfights of the First World War, where honour was paramount and combatants were expected to fight fairly, were quickly dashed. The clashes of these metal monsters were brutal and violent and sides were usually unevenly matched.

Many writers have sought to immortalise the honour and gallantry of the German Panzers by branding them 'Knights in Chariots'. But, in those first, bloody encounters with them, Doug saw nothing chivalrous about these merciless killers. Their blitzkrieg tactics relied on overwhelming force and constant motion coupled with sustained and accurate fire.

With heavy machine guns, flame throwers, rocket grenades, big guns and the ability to grind the infantry under their churning tracks, they became ruthless harbingers of death in its most gruesome form, where victims were either blown to smithereens or roasted to death.

The latest Axis push, Operation Oxhead – led by Tigers and Panzers – had begun at a tiny rail head called Sidi Nsir. Rommel's intention was to sweep down the valley and take

Hunt's Gap, a British-defended defile east of Ksar Mezouar which guarded the approach to Beja and the west.

At Sidi Nsir, the Royal Hampshire Regiment's 5th Battalion and the 155 Battery of the Royal Artillery faced an attack of overwhelming strength but were ordered to hold their positions at any cost. It was a formula for absolute carnage.

Right from the outset, the men had little or no chance of survival. In dreadful weather conditions, constantly strafed by Messerschmitt fighters, they fought almost to the last man.

At dusk, when the British were finally permitted to withdraw, only 9 out of 130 artillerymen remained alive. Of three companies of the Hampshires, two had been wiped out and only thirty men of the third company survived to escape or be taken prisoner.

They had succeeded in delaying the enemy advance but it was achieved at a terrible cost in lives. However, their sacrifice had bought time to move reinforcements into Hunt's Gap.

At a briefing by Brigadier Richard Maxwell on 27 February at 25th Brigade field headquarters, Doug learned that the Allied forces were bracing themselves for a follow-up by the Germans the next day. They were thought to be planning a massive thrust north of Hunt's Gap.

The North Irish Horse, a tank regiment that had sailed out with Doug from England, would take the brunt of the anticipated attack, backed up by five batteries of field artillery and a strong detachment of infantry. The Royal Engineers would also be there in force, as would Lieutenant Reg Whatley and his brigade recovery unit.

Doug met with Reg at first light and in time to see one of the Irish Horse tanks being knocked out by a German anti-tank gun, killing three of the crew instantly.

The rest of the crew barely had time to bail out before Reg

sped forward in his recovery tractor to take the stricken vehicle in tow. It was clear to Doug that the brave young Lieutenant had only one way of doing anything: by example. Leading the way into the most extreme situations and never asking or telling his men to do anything he wasn't prepared to do himself became, from the off, Whatley's standard way of operating.

The battle raged for three days around Ksar Mezouar station. Under constant enemy fire, Doug, Reg and their teams were pushed almost to the point of exhaustion while dragging broken-down or damaged Churchills out of the field to the temporary repair shops, where dozens of engineers worked around the clock to get them back in action. Twice the Allies were overrun – and twice they beat the Germans back.

The two men kept in touch by radio. Despite the critical nature of the work they were doing, Doug had not for a moment forgotten the real purpose for his being there – to capture a Tiger. If the opportunity presented itself, then they had agreed to join forces and swoop in together.

But, in their running battles with the Allies, the Germans were keeping the big Tigers at a frustrating distance where they were less likely to come in range of the main British gun batteries.

During the first afternoon of fighting, one of the Tigers had momentarily strayed towards the front of the battle zone, and was holed through its turret by one of the Irish Horse Churchill tanks, whose commander then courageously, and perhaps over-optimistically, raced forward to try to finish it off. Doug saw what was happening and raised Lieutenant Whatley on the radio.

'I think there's a Tiger in trouble,' he yelled into his microphone, then groaned in the next breath as the Churchill broke off its attack. Its commander had been struck in the back of the neck by a bullet – or shrapnel from the shells exploding all around him – and his number two had decided to rush him back to a casualty station.

The Tiger, which had paused for a few seconds, now appeared to be unaffected by the Churchill's shell as it retraced its path to disappear behind the hill it had emerged from.

'As you were,' said Doug dismally. 'He's gone back into hiding.'

It was not until the next afternoon that another Tiger was spotted, this time by Reg Whatley. For reasons unknown, it appeared to have been abandoned by the enemy along with a handful of Panzer IIIs and IVs. From a distance of several miles, he judged it to be basically intact. There were other enemy tanks that had been set on fire or destroyed by a relentless bombardment from British anti-tank guns, artillery batteries and the canons of the Irish Horse, assisted by a squadron of RAF dive-bombers which had kept up a non-stop pounding of the German lines.

Doug had just returned to his main workshop, several miles behind the front, when he received the news from Reg. Without hesitation, he instructed the Lieutenant to meet him at a point near Ksar Mezouar station with his recovery tractor. After quickly rounding up Shaw and Rider, he signed out a Jeep and sped as fast as he could across the muddy ground to his rendezvous.

He had expected to receive a jubilant welcome from the Lieutenant, who he knew to be as excited as himself at the prospect of getting his hands on a Tiger, but instead he found Reg with shoulders slumped and downturned mouth sitting on the step of his tractor, giving a thumbs-down signal with his right fist.

'What on earth's the matter?' he asked. 'Somebody killed?'

Reg shook his head. 'Not somebody. Something. I've just heard that our Tiger's been well and truly killed off by the Royal Engineers.' His usually warm baritone voice had a sad, plaintiff note.

'What?' Doug looked crestfallen.

'Yes. Afraid so. The CO sent a party of engineers forward with instructions to destroy all the enemy tanks that had been abandoned. They packed them inside and under with high explosives and blew the whole bloody lot to pieces, including our Tiger.'

'Are you absolutely certain?'

'Yes, sir. That's what they say.'

Doug felt like he'd been punched in the belly. It was one thing to have the enemy blow up one of its own tanks to thwart him. But for his own side to do it was very hard to come to terms with. Still, Doug wasn't ready to give up yet. He had to be certain all was not lost.

'I think we should go up there and have a look for ourselves. Just in case they've got it wrong. We'll take your tractor. It'll handle the mud better than this Jeep.'

He turned to his men. 'You, Sergeant Shaw, stay here with the Jeep. I don't know how long we'll be. But one way or another we'll need transport back to the base. Rider, you come with us.'

It was almost nightfall when they reached the site of the latest tank graveyard. But there was still enough light to discern that the Tiger was beyond salvage. The turret and interior were shattered by an internal charge and a second charge underneath had destroyed the tracks and bogies and even split the hull.

Doug shielded a cigarette with his cupped hands and lit it, walking a few yards away from the wreckage.

'I think if we see a Tiger in trouble again and there's a genuine chance of capturing it then we're going to need to move in quickly, and that means we'll need a bit more than a tractor for protection. It's dangerous enough doing what we already do. But chasing a Tiger is a damned sight more hazardous.'

'Perhaps we should use a tank ourselves, sir,' said Reg. 'One that we've repaired but isn't quite good enough for full active duty.'

Doug pondered for a moment, taking a deep drag on his cigarette, and his face lit up. 'I do believe you've put your finger on it, Lieutenant. That's what I'll use in future. It can be used to tow our chaps out of trouble just as efficiently as a tractor.'

He called over to Rider, who had wandered over to examine one of the battered Panzer IVs, 'Come along, Corporal. Time to go home.'

'Right, sir,' said the NCO, throwing the stub of his cigarette in the mud.

Rider strode across to join them. As he walked, he heard a loud click from underfoot and froze to the spot.

'I've trodden on a mine,' he yelped. 'Get down.'

Corporal Rider threw himself face down in the mud. Forty yards away, Doug and Reg reacted instantly and followed him down, bodies spread-eagled in strict compliance with the instructions of the Sapper Sergeant who had lectured them on the minefields they would encounter in their work. He had put the fear of God into all of them, and now it was paying off.

A fraction of a second later, the charge from the mine leaped two feet into the air and exploded, sending metal ball bearings firing horizontally in every direction. These mines were designed to injure rather than to kill and could maim a soldier up to a hundred yards away in a 360-degree radius. An injured soldier was of more value to the enemy than a dead one. It sapped your opponent's resources more to evacuate him from the battlefield and deal with his wounds and post-surgical care than to deal with a corpse.

The only defence against these mines, apart from being in an armed carrier, was to do what Doug and his companions had done – hug the ground, letting the ball bearings whizz past over-head.

Now, as they scrambled to their feet to examine themselves and each other for injuries, Doug began to laugh. 'We look as if we've been involved in a mud-wrestling contest.' Hardly a patch of skin was visible on any of them. 'But thank God for Corporal Rider's timely warning. You almost certainly saved our lives, Rider. I'll make sure it's mentioned in my report.'

'Another reason to do our hunting in a tank,' Reg commented, wiping mud from his hair. 'It may not be a whole lot safer but, gosh, it's surely a hell of a lot cleaner.'

The rest of March saw almost continuous action for them all, with the Tigers making an occasional, and usually devastating, appearance, knocking out the Allied tanks at a score rate of more than six to one.

Reg and Doug saw Tigers temporarily put out of action on two more occasions. They were beaten to the first by German engineers who wasted no time in towing it from the battlefield, almost from under their noses. The second Tiger was destroyed by the abandoning crew using high explosives. By the time Doug and his unit reached it, all that remained was a pile of useless scrap metal.

As March ran into April, Doug became increasingly frustrated. It was nearly three months since Churchill had dispatched him on this mission, and, though he knew it was irrational, he blamed himself for his failure.

'We've simply got to come up with something,' he told Reg as they shared a drink in the makeshift mess tent on Doug's birthday.

Reg raised his glass. 'Perhaps you're getting too old for it,' he said mischievously. 'Many happy returns, sir. I'm sure we'll catch your Tiger soon.'

Doug was just 29 years of age.

CHAPTER FOURTEEN
21 APRIL 1943

Doug Lidderdale awoke in his tent in the British Base 104 REME workshop and instantly knew something was wrong. The shuddering howl of a wild mountain dog, close to the camp, filled the night, but instinctively he knew it was not the sound of animals that had awoken him.

He swung his legs from the camp bed to the slatted wooden floor and groped at the top of the rickety bedside cabinet for his torch. It threw a feeble beam. He reminded himself to ask AQMS Shaw to draw new torch batteries from the stores later in the day. He threw a little light on his watch, which showed 0500 hours. He now recognised the sound that had awakened him – heavy gunfire. He could tell it was several miles away. He turned his shoes upside down and vigorously shook them to check for scorpions. Then he slipped them on, pulling the laces tight but not bothering to tie them properly. He stood and jerked back the flaps of his tent.

Even in his full desert uniform – in which he had slept – the temperature felt barely above freezing.

'Either it's swelteringly hot or brass-monkey cold. What a

bloody country,' he muttered as he stared in the direction of the gunfire which was coming from the northwest – a location dubbed by some wit within the British infantry as 'Peter's Corner,' because of the likelihood of meeting St Peter there.

The night was crystal clear and the sky was studded with shimmering stars. But much brighter than even the biggest stars were the flashes beyond the nearest hills, which were almost continuous, sure signs of a full-scale battle going on. And yet Doug knew that Operation Vulcan was not scheduled to start for another 24 hours.

'What the hell's going on?' he asked himself.

Vulcan was General Alexander's end game. Knowing he had 250,000 Axis troops bottled up in the top northeast segment of Tunisia, Alexander intended to force surrender, or sweep them into the sea. In Casablanca, he had promised Churchill a successful conclusion of the war in North Africa by mid-May, and he intended to keep that promise. He had made that quite clear at a meeting Doug had attended, representing the field engineers – the Desert Rats – in March.

But Doug surmised Germany's General Hans-Jurgen von Arnim had a different agenda. He concluded correctly that the Afrikakorps, the German expeditionary force in Libya and Tunisia during Erwin Rommel's North African Campaign, must have got wind of Alexander's plans and was attempting to out-fox him by launching an offensive of their own when it was least expected.

'Morning, sir. Do you happen to know what the hell's going on?'

Doug recognised the voice and could just make out the slim shape of Lieutenant Reg Whatley who had a torch cupped close to his side.

'Morning, Reg. I don't know for sure, but it's my guess the Krauts are intent on screwing Vulcan before it gets started. They're definitely on the move. By the sound of it, some of our chaps might be getting one hell of a pounding. One thing's for certain, there are tanks out there, a lot of them. And you can bet your life that in among them there are Tigers on the prowl. This could be our lucky day: the chance to nab one at last.'

'Do you really think so, sir?' said Reg, his voice hoarse with excitement.

'I do. Get the lads together, grab a bite to eat and be ready to roll the moment the call to action comes. As I'm sure it will. '

'I'll personally make sure Bessie is all fuelled up and ready to go the moment you need her.'

A pained expression clouded Doug's face. 'Reg, do we really have to call a bloody 40-ton Churchill Mark Four by such a ridiculous female name?'

'It's not my idea, sir. It's Sergeant Shaw,' said Reg apologetically. 'That's the name of his mother-in-law who, he swears, is built like a tank. It's the only way he can have a go at her without his wife knowing about it.'

Doug laughed. 'Well, I suppose if he's prepared to risk his neck trying to catch a Tiger by the tail, then the least we can do is help him take the piss out of his wife's mother. Bessie it is then. But, for God's sake, don't let any of our brother officers in tanks get to hear of it.'

Reg hurried away. Doug shone the weak beam of his torch on the ground ahead of him every few paces. He collected his towel and shaving kit from the tent and then walked carefully across to the officers' showers, open at the top and screened off by a six-foot canvas fence.

The water, heated by the North African sun during the previous day, had already cooled to below body temperature.

Doug was shivering when he stepped from the shower and vigorously rubbed himself dry with the rough army-issue towel. He shaved by the light of a shielded hurricane lantern. Doug fondly remembered the last hot shower he had shared with Kate in England and for a moment his thoughts carried him 2,000 miles from war-torn Tunisia to a genteel hotel room and the last passionate hours spent with his gorgeous blonde bombshell of a wife.

He had been granted only two nights, courtesy of the Prime Minister, to familiarise himself with Kate's naked body – but he found he could visualise every tiniest curve and love spot.

He tried to imagine what her reaction would be to his going out today in the hope of capturing one of the near-invincible German Tigers. He chuckled to himself. She probably would have said something about boys and their toys, he thought; Doug and his gang playing Just William – but with far more at stake than the unsought adoration of Violet Elizabeth Bott.

Grinning at his private thoughts, he returned his kit to his tent and went in search of breakfast. By now, the mess tent was almost visible and the eastern horizon, still showing the distant flashes of heavy artillery, was beginning to lighten from inky blackness to a dull-grey orange.

A recently promoted REME Captain materialised from the dimly lit mess tent with a clipboard under his arm. He saluted smartly. It was Doug's 2.I.C – his second in command.

'Morning, sir. I was just coming to wake you. According to HQ, Jerry has launched a full-scale attack on Banana Hill about five miles east of Medjez. Tanks, infantry and heavy artillery.'

Doug listened and said, 'We're well dug in up there.'

'Well, sir...'

'Out with it, man.'

'Our own artillery and the Duke of Wellington's Regiment

weren't ready for them and they have been more or less overrun. Details are sketchy at the moment but a good many vehicles on our side have been blown to bits or set on fire, and casualties are probably high.'

'Just how close are they, Captain?'

'From what I've gathered, sir, the enemy have swung south towards the Goubellat road. Brigadier Colvin, who's in charge of the 24 Guards Brigade there, has stood firm and called in a squadron of Churchills. He's holding Jerry off. The Duke of Wellington's have regrouped and have opened fire from the high ground in support.'

'What about nearer to home? Is this a push all along the front?'

'Again, the intelligence is sketchy, sir, but HQ seems to think so. They've ordered another battalion of the Royal Tank Regiment up from their harbour at Testour. They're expected to be in our area by mid-morning. One squadron's been put under the 21st Tank Brigade whose HQ is only a couple of miles in front of Jerry's main thrust, and two under the 10th Infantry Brigade and Brigadier Hogshaw.'

'Then they'll be standing alongside us, is that right?'

'Yes, sir. Though they haven't been blooded yet, sir. This will be their first taste of action.'

'Thank you, Captain,' said Doug. 'You can be damned certain that blooded or not they'll be up to the job. This is what they've been training for. Now they have the opportunity to give the Hun a damned good pasting, you can bet your life they'll give it all they've got. A baptism of fire it may be, but I know we can rely on them to give the enemy a taste of hell. Remember, Captain, it's our job to make sure they have the best support we can give them. That's what we Desert Rats are here for.'

'Yes, sir.'

'When the balloon goes up, I want one recce scout car near the front to keep us informed, and everyone and every recovery vehicle ready to retrieve damaged tanks and other vehicles. Then they must get them repaired and turned around in the fastest time possible.'

The Captain nodded and said, 'Will you be going out yourself, sir?'

'Yes, I will. And you'll be my wingman. We'll liaise by radio. But first of all let's make damned sure all the men are fed. It may be the last chance they have to eat before nightfall. If Jerry's push is as strong as it sounds, they'll probably have to fight through the night as well. There's bound to be one hell of a lot of damage. Let's just pray we can do worse to them than they do to us.'

'I'll join you in that prayer, sir,' said the Captain. He saluted once more and marched away smartly.

CHAPTER FIFTEEN
21 APRIL 1943 – MORNING

At the same moment as Douglas Lidderdale was about to eat breakfast, Major August Seidensticker, one of Germany's legendary Panzer heroes, was addressing his remaining platoon leaders and their tank commanders eight miles to the north, outside Montarnaud in Tunisia.

The dashing 38-year-old from Dusseldorf was commander of the Schwere Panzer-Abteilung 504 (Heavy Tank Battalion 504), which had incorporated the remnants of the Afrikakorps' s.Pz.Abt.501.

Seidensticker, a tall, deeply tanned Panzer veteran with a strong athletic build and a personality to match, had been a career officer since the age of 20. He had seen action in Poland, where he had earned the Iron Cross – both first and second class – and had survived the western campaign and the Russian front, where he had added the German Cross in Gold to his awards, before being shipped with his 18 Tigers and 25 PzKpfws, just three weeks ago, to Tunisia, where he and his men were expected to perform miracles. Seidensticker was the genuine article, an authentic German hero tried and tested. And utterly disillusioned with his High Command.

Hitler had refused all pleas from both Rommel and the new German Commander-in-Chief, von Arnim, to evacuate their forces to Sicily or Italy and had issued orders that every man should fight to the death: 'Conquest is our purpose. Not retreat. Surrender should not even be contemplated as an option.'

General von Arnim had personally confided that today's attack on an enemy that vastly outnumbered them with triple their number of tanks and artillery, and whose combined air forces dwarfed the tiny remaining squadrons of Luftwaffe planes in the theatre, was a quixotic, lunatic and almost certainly suicidal gesture to please a pair of mad dreamweavers in Berlin and Rome.

Il Supremo, the Italian head of the Axis forces in North Africa, safely billeted in Rome, had his own madman to deal with. Mussolini retained an obsessive belief in Nazi might. He clung to his vision of a modern Roman empire stretching from the Suez Canal to the Atlantic, and had instructed Il Supremo to secure Tunisia and claw back Egypt and Algeria at any cost.

'We must prevail,' he ordered. 'It is our destiny.'

While this rhetoric gushed forth, no attempt was made to ship desperately needed supplies of ammunition, fuel and food to his armies. Seidensticker knew some units of the surviving Italian forces were literally facing starvation, and many Italian infantrymen were left with barely more than 50 rounds of ammunition per man.

Now, in a crazy but glorious gesture to their grotesquely clown-like leaders, von Arnim launched a grandstand finish; one last, magnificent offensive against the Allies, aimed at the centre of their eastern front. He had code-named it Operation Fliederblüte (Lilac Blossom) after his favourite springtime bloom in his native Silesia.

Paratroops of the Hermann Goering Division would lead the attack, backed by a formidable tank force led by Major Seidensticker. Three platoons, each consisting of two Tigers and a support group of PzKpfw IIIs and IVs had been committed since before dawn. Now Major Seidensticker himself intended to lead his remaining two platoons in the main German thrust against the British. Von Arnim was contemptuous of the Americans to the north and the French to the south. The real strength, he maintained, lay with the British. Break them, he had laughingly told von Arnim, and there might still be some mileage left in this damned and doomed campaign. But it was hollow laughter and both men knew that it was no more than a forlorn fantasy.

The Panzer ace was keenly aware that, in this final briefing before the battle, it was now his task and duty to convince his men that they were not simply embarking on what he knew to be a useless, suicidal venture, but on a glorious mission which could still snatch victory from defeat for their beloved fatherland.

He clapped his hands, calling the men to order, and then, after removing the glove on his right hand, he pointed one dramatic finger towards the eastern horizon, where, heralding the day, the golden orange of the rising sun sent yellow, red and pale-blue horizontal streaks shooting across the craggy landscape.

He gave a rallying speech to his men. 'Gentlemen, by the time this sun sets on us tonight, we will be enjoying the bars and brothels of Medjez.'

This drew the expected laughter, a few raucous comments and a huge cheer from the assembled troops.

Seidensticker continued, 'I can't promise you it's going to be easy.'

'That'll be a first,' yelled one young officer, to further laughter.

Seidensticker grinned. 'But I can promise you there'll be plenty of targets for everyone, and you all know that knocking over these Churchill tanks is like shooting ducks on a fairground stall.'

More laughter and a further cheer.

'But remember, it's their artillery that can do the damage, to us and to the infantry. Make their six-pounders your principal targets. Knock those out and we can pick off their tanks at our leisure. Our own artillery is already dug in on the hills to the east of the Goubellat road and our infantrymen are ready to advance. Other units are already engaged with the enemy further to the north and I am assured we are making good headway towards Medjez.

'It's down to us to pincer in from the south and combine with the rest to make a decisive strike on the town. That, gentlemen, is our goal. That is what I expect you to achieve. We leave here in the next hour and I want all platoons to be in place by noon.

'All of you know that we've been killing up to ten of their tanks to every one of ours they bag. Today, I say to you, let's push that ratio even further in our favour. The German tank corps is the finest in the world and you men are the finest of the finest. That's why you are here and why we have been entrusted with this sacred mission.

'Yesterday was our Fuhrer's birthday. Today, you give him a great present by fighting like never before. You fight for the Fuhrer, the fatherland and the future. Know also that it is an honour for me to serve with you, and I salute you.'

Instead of the usual stiff-armed Nazi salute, Seidensticker raised his right hand smartly to his forehead and gave his men the traditional Panzer salute. Every man snapped his heels together and returned his salute in kind.

'And may God help you,' the Major muttered to himself as his men, many of them barely in their twenties, eagerly dispersed to join their waiting tanks.

'For your beloved Fuhrer has already written you off,' he added sadly, with a resigned shake of his head.

The beloved Fuhrer had enjoyed his 54th birthday at the Berghof chalet in the Obersalzberg of the Bavarian Alps near Berchtesgaden. In 1938, Hitler had written in the British *Homes and Gardens* magazine, 'This house is mine. I paid for it from the money I earned as a writer.' His vegetarian diet was supplied from the kitchen gardens, and guests were only allowed to smoke on the terrace. On this day, 20 April 1943, Eva Braun and her close friend Herta Schneider entertained Herr Hitler with songs by Irving Berlin and Cole Porter. Adolf played the piano and, as usual, his German Shepherd Blondi fought with Negus, Eva's Scottish terrier. At night, Hitler and Eva Braun went to their respective bedroom suites with the interconnecting doors. It was just another quiet day in the life of the most evil dictator the world had ever known.

CHAPTER SIXTEEN
21 APRIL 1943 – AFTERNOON

Beyond the hills to their west, the members of the Royal Tank Regiment's 48th Battalion – many just as young and eager as their German counterparts – were manoeuvring their Churchill tanks into the designated positions, keen to put their months of training into practice.

On both sides, each man was secretly fearful of the coming clash. Only someone who was totally insane could enter a battle without realising that many were facing violent death or hideous maiming and disfigurement. Most would hide their fear and bravely confront the risk, because that was what was expected of them. These were mostly ordinary men doing an extraordinary job, laying their lives on the line for their countries and the honour of their regiments.

Unlike the great battles of history where huge armies met face-to-face and looked the enemy in the eye, today's warriors fought often at a distance along a constantly changing front. Sudden thrusts and retreats were common, the commanders moving their men around like chess pieces in a game of bluff and dare, with the ultimate checkmate – death – waiting for anyone who made a wrong move.

The Churchill tank which Doug had commandeered – a battered veteran of earlier scraps that had been repaired in the 104 workshop – reached the southern edge of the intended battlefield by mid-afternoon. Ahead of him, alien to an English eye, stretched a vast area of hills and scrub and rocky ground and, surprisingly, a few large fields of corn.

From his standing position in the commander's cupola he could see a ridge of hills rising to his right, which the Germans had seized during the night. They were already well installed, with artillery dug in along the escarpment hillside. Below his position and a mile to the north, he could make out several British tanks scattered across a series of cornfields and deep scrub. They were slightly to the west of two wadis, the one furthest from them being steeper sided and deeper than the other.

A second squadron was northeast of this position – partly concealed behind a ridge, held in reserve.

Brigadier Hogshaw's plan was to mount a counter-offensive and attempt to dislodge the Germans from two small hills defended by sections of a parachute regiment and artillery in front of their main position. The job of actually taking these hills was assigned to the 1/6th East Surrey Infantry supported by the tanks.

Later, Doug would discover that a major fault in this plan was the complete lack of wireless communication between the tanks and the infantry advanced headquarters. He was stunned at the oversight – there was no liaison between the two attacking forces. It should have been arranged at the 10th Infantry Brigade's headquarters meeting that morning, but the meeting was dramatically cut short when enemy artillery fire scored a bull's-eye on the HQ. By the time it was safe to reconvene, the infantry commanders had returned to their units and no wireless link was ever agreed with the tank battalion.

But Doug knew nothing of this when, scanning the ground ahead, he saw the leading members of the East Surreys cross the first wadi and begin their approach to the Mehirigar, the nearest hill – or Djebel – ahead and to his right. They appeared to be attacking without any cover from the tanks, which were half a mile north of them and, Doug realised with alarm, seemed to have become bogged down in the deeper of the two wadis before their advance had properly started.

One of the Churchills had plunged nose first into the wadi, and was stuck in the sand, its six-pounder gun half-buried. By the time another tank had been manoeuvred into place to pull the partly buried lead behemoth free, it was almost 1500 hours and the East Surreys, without the promised support, were halfway up the slopes of their objective and exchanging heated fire with the enemy.

Lieutenant Reg Whatley, seated on the hull roof next to Doug, had his binoculars trained on the distant tanks. 'They seem to be having a tough time getting across that old riverbed, sir. Should we lend a hand?'

'No, Reg. That's not why we're here. Anyway, they seem to have found a way round now. It's obvious to me that somebody didn't reconnoitre their path in advance. Christ, if the enemy's artillery or tanks had been aware of it, there could have been absolute carnage. They were sitting ducks there for nearly half an hour. We could have lost half our Churchills and their crews.'

They watched with some relief as tanks of the leading troop rolled across the cornfields and began climbing the stony slope, pounding the Djebel el Mehirigar ahead of the advancing infantry, and finally providing them with much needed backup.

At that juncture, the main enemy artillery on the hills to the east opened fire, using 50mm and 75mm shells to concentrate their initial efforts on the tanks. Moments later, the first German tanks lumbered over the low ridge between the wadis and the Germans' main front, and instantly the British artillery six-pounders, dug in behind the strike forces, added their voice to the bedlam.

Training is all very well, thought Doug, but there is one thing about a real fire-fight for which no amount of training could prepare you: the volume of war – the relentless head-pounding din which drained your senses.

'God, sir, I've heard of all hell being let loose but I never expected to be sitting there when it happened,' said Reg. 'I feel bloody sorry for the infantry, caught out in the open. They haven't even got trenches to shelter in.'

'I don't know about that,' replied Doug, 'they don't look as if they're feeling sorry for themselves. In fact, I'd say the exact opposite. They seem to have flushed Jerry off that first hill and taken control.'

He leaned forward, binoculars to his eyes. 'Tally-ho, Reggie!' Doug shouted above the noise.

Whatley cupped a hand to his ear. 'What, sir?'

'There, Reg – straight ahead – and making those Panzer IIIs and IVs look like Dinky toys. A pair of Tigers.'

Reg's face lit up. 'That's wonderful, sir.' Then he threw Doug a quizzical look. 'But how exactly do we plan to bag one?'

Doug grinned. 'I don't know yet. All I can hope is that our boys inflict some mortal damage and enable us to go in and commandeer the remains. It's a question of waiting and seeing what happens and then seizing any opportunity which presents itself.'

He was drowned out by another explosion. A Panzer III had

exploded, splitting the hull. Doug watched a shattered body hurtled skywards from the open turret in a gush of blood, fire and smoke.

'Jesus, Reg, did you see that? First kill to us.'

A German tank had rounded a small hillock and virtually ran into Captain Alan Lott, leading the three Churchill tanks in the third troop of the 48th Battalion's 'A' Squadron, which he commanded. Doug had been introduced to the young tank officer with the engaging smile, film-star looks and dark wavy hair, and was impressed with his maturity and quiet air of confidence. Now, although he would not learn the officer's identity until later, he witnessed that confidence in action and was again impressed. Involved in his first real engagement, and within a second or two of being confronted by an enemy tank, Captain Lott had brought his own gun to bear and fired – the recoil rocking the 40-ton tank back on its treads. The six-pound shell scored a direct hit, penetrating his opponent's hull and causing the explosion and fire that decimated him.

Scarcely before he had time to register his first taste of real action and his first kill, never mind celebrate his victory, disaster struck. Doug found himself crying out aloud in horror as he watched the young Captain's own tank explode in a thunderball of flames and flying metal. A German shell – fired, he believed, from one of the Tigers – penetrated the turret just below the Churchill's six-pounder gun, behind which was the Captain's cupola.

Captain Lott, his uniform and hair on fire, and his wireless operator were able to leap from the turret before the whole interior compartment became an inferno. The driver, co-driver and gunner were either killed instantly by the initial explosion or burned to death – the nightmare fate feared by all tank crews.

Reeling from the shock of seeing a British tank and most of its crew obliterated in less time than it had taken him to draw one breath, Doug found himself holding that breath as he watched helplessly and saw the nearest Tiger open fire again. The massive tank shuddered under its own recoil.

It was a devastating shot. The 88mm shell passed straight through the hull of the tank commanded by Lieutenant Peter Gudgin, leader of 4 Troop, who was standing on the commander's pedestal. The shell entered through the Besa machine-gun mounting and passed between Gudgin and his wireless operator before slamming into the engine, which immediately caught fire. Later, Gudgin would recall to Doug that he felt the shell pass his right leg. It was that close.

Gudgin ordered his crew to bail out, but before they were clear their tank was struck twice more, once on the right of the turret by a 50mm shell, which lodged in the armour, and once by a 75mm artillery round, which bounced off the left side of the hull.

Gudgin realised immediately that damage from the Tiger's shell had somehow caused the driver, Lance Corporal Bob Fletcher, and the tank's co-driver, to become trapped in the forward section of the hull. Without hesitation, he and his wireless operator dived back in through the turret hatch to help haul them free, while a German heavy machine-gun nest on the hillside half a mile away raked the stricken Churchill.

They dragged their fellow crewmen to safety and scrambled from the turret under a hail of bullets. With nowhere to hide, they would almost certainly have perished had it not been for an incredible act of bravery by Gudgin's Troop Sergeant, 'Plum' Warner. Seeing their near-hopeless situation, he put his own life on the line – and those of his crew – by positioning his tank as a shield between the enemy fire and his comrades.

But, instead of taking advantage of Warner's cover and making their escape, Gudgin and his crew chose the far more hazardous option of zigzagging their way across the rocky ground to where Captain Lott – still alive but dreadfully burned and in agony – was attempting to rescue the three missing members of his own crew. It was a hopeless task, as, by now, in addition to the roaring flames pouring from the open turret, some of the tank's ammunition had begun to explode in the compartment where the doomed trio was entombed.

It was inconceivable that anyone could still be alive in that inferno, but Captain Lott and his wireless operator refused to accept this, or listen to reason. Gudgin finally took the tough decision to leave him there. With profound reluctance, he turned away to lead his crew across a minefield via abandoned enemy trenches to the safety of their own lines.

Two weeks later, Captain Lott and his radio man died in hospital from their injuries.

Doug had watched this sorry saga unfold from his position on the commander's pedestal of his tank behind a low ridge almost totally obscured from the enemy's view. The sheer size and ferocity of the Tigers was quite terrifying. He was certain some of the Churchills had scored hits on the Tigers but their shells had seemed to simply bounce off, inflicting no apparent damage at all.

During the next hour, from his vantage point on the sidelines, Doug saw two more Churchills put out of action as well as two Panzer Mark IIIs and a Mark IV, though only the Mark III appeared to have been completely destroyed.

It was also plain to Doug that the British tanks had no idea what was happening to the East Surrey Infantry who, having dislodged the Germans from their newly taken outpost on the nearest hill, had run into strong resistance in their attempt to cross a low ridge and take their second objective – the higher hill Djebel Djaffa – a mile to the northeast. Doug could see that the men of the Hermann Goering Jaeger Regiment and Grenadiers outnumbered the East Surrey infantrymen by about two to one and were laying down a withering hail of machine-gun and rifle fire and grenades. He watched through his binoculars, almost in disbelief, as one British officer led a platoon of men with fixed bayonets at a German machine-gun nest. Had it been in his power, he would have presented every one of them with a medal on the spot.

These incredibly brave soldiers were under constant and heavy enemy fire without proper tank support. Doug knew that their position was becoming dire.

The Churchills were concentrating all their attention on the German tanks and heavy artillery. Their primary role – support for the infantry – seemed to have been forgotten about, ignored or considered irrelevant under present circumstances.

In the control chamber of a tank, temperatures reached up to 120 degrees. To escape it, Doug's entire crew had positioned themselves atop either the hull roof or the turret, brushing off clouds of flies and midges, while observing the distant battle. The consensus was that, while the British had taken their first objective through sheer guts and determination, they were unlikely to be able to hold on to it. The Germans were well dug in and were significantly superior in numbers to decide the outcome.

'It seems like stalemate at the moment,' said Reg. 'Man for man, though, I think our chaps are winners all the way.'

'Except for their bloody Tigers,' replied Sergeant Shaw. 'Nothing we throw at them seems to even dent them. If they had a few more of them, I reckon we'd be in truly dire straits.

'Beggin' your pardon, sir, if I may ask what may seem a pretty obvious question, but how can you expect to take one of those monsters with what we're sitting on? It would be like Lieutenant Whatley here going up against Joe Louis. With respect, Mr Whatley, he'd bloody kill you. Same as one of them Tigers would do to us. We need us a giant-killer, sir. That's what we need.'

Doug smiled briefly at Shaw's graphic analogy. 'Going head-to-head with a Tiger has never been an option, Sergeant. But, if someone else gets to maul him first and pulls some of his teeth, that's where we come in.

'We know it's happened before because we've seen it. But, if you remember, Jerry managed to drag one of the corpses away before we could get to it. They put a big bundle of dynamite under the other one after stripping it down. It was useless to us.

'We've just got to hang around and wait for one of our chaps to hit one of 'em where it hurts and hope we can get our hands on it before Jerry does. It'll be getting dark soon so we may have to come back tomorrow or the next day. But at some point we'll get lucky. So stay sharp and keep your eyes on those Tigers.'

Doug glanced at his wristwatch. 'It's gone five o'clock. We'll give it one more hour and then we'll call it a day. Perhaps it's time to pass the water round. Reg, you and Corporal Rider keep an eye on the Tigers. Sergeant Shaw, break out the water, while I pay a call of nature.'

Doug climbed out of the cupola and carefully down the Churchill's side, trying not to touch his bare arms and legs against the hot metal.

Doug had relieved himself and was remounting the hull when Lieutenant Whatley called him, his voice full of excitement. 'Sir, come quickly. It looks like one of the Tigers is in trouble.'

Doug followed Reg's pointing finger and focused his binoculars.

Less than half a mile away, beyond a low ridge, he could make out the turret of a stationary Mark VI Tiger. The cupola's hatch was thrown back and a man was leaning out and seemed to be examining the gun.

Doug realised that this was the commander.

His order came low and urgent. 'Everyone get into place. Start up the engine. This just might be our lucky day.'

He could see from the exhaust gases that the Tiger's engine was still running, but there appeared to be a problem. Doug watched the gun as it was elevated then lowered. The commander was crouched down in his cupola and was evidently struggling with something. Suddenly, he pulled himself clear of the hatch and crawled forward towards the base of the gun, which was covered with a protective mantle, and carefully examined the edge of the turret.

A small hatch towards the front of the turret opened and the head and shoulders of another German crew-member appeared. He too seemed to be looking at the base of the gun and having an animated conversation with his commander.

From Doug's perspective, it was like watching a scene acted in mime. Then, with an intuitive flash of inspiration, he realised what the problem was.

'The bloody turret won't turn,' he screamed to Reg. 'Something has happened to his turret mechanism. It won't bloody turn. This is what we've been waiting for. Corporal Rider, get back into the driving seat and give me full speed. Now.'

He scanned the ground ahead. 'We'll cut around the end of this ridge and then down the slope so he won't see us coming. If we're really lucky and this bloody gunfire keeps up, then he won't hear us either. At top speed, we can be right up his backside in just a few minutes.'

Doug fed instructions to Corporal Rider, who was already at the controls. They trundled forwards, churning up clouds of dust as the treads gained traction. They cleared the end of the ridge, still out of sight of their quarry, and headed across the slightly depressed area between themselves and the ridge the Tiger lay behind.

Doug kept checking to the northeast, where the main German forces were entrenched in the higher range of hills. He prayed that another German tank would not spot them and become involved.

He could hardly contain his excitement. Adrenaline pumped through his veins – the fever of the hunt had him totally in its grip. He was aware of the danger and that he and his men were careering at top speed into an extremely precarious battlefield situation where men on both sides were being killed and mutilated, but he felt no fear, just an overwhelming desire to get to grips with the enemy and secure his prize. Doug had risked his life before, rescuing the Brigade's damaged tanks in the midst of battle. But this was different.

This time his objective was an enemy tank. His quarry.

His Tiger.

He had even been able to make out the number: Tiger 131.

Doug felt something in himself that he had rarely experienced before. Pure, unadulterated excitement. He felt truly alive.

The Churchill hit the slope and climbed without difficulty towards the turn of the ridge using its one, huge advantage over its German tank rivals – the ability to effortlessly conquer steep gradients.

'Reggie, pass me up a Sten gun and some spare ammo,' Doug shouted down into the hull. 'Then get yourself armed and tell Shaw to arm up and come here to cover me. We are going to get in close. So close they can't use their cannon on us. I reckon we'll make contact with Jerry in about a minute from now.'

Doug leaned forward, eager to get his first close-up sight of his quarry. The hot desert breeze was blowing sand in his face, but he didn't notice. It was an exhilarating moment – knowingly entering the danger zone but completely unafraid. He had become the hunter and now all that mattered was the kill.

He sucked in a breath. 'Let's do it,' he yelled.

CHAPTER SEVENTEEN
21 APRIL 1943 – LATER

On PzKpfw VI number 131, a commander – an Untersturmführer – had a tough decision to make. A lucky shot, either from one of the British Churchill tanks or from their artillery backup, had struck the Tiger's turret ring, making a slight gouge. The damage to the armour was negligible, but the effect was catastrophic, as somehow the shell had lodged in the turret's revolving mechanism.

His gunner could still manually elevate and depress the 16-foot 88mm gun but neither the hydraulic traversing mechanism nor the manual traversing hand wheel could make the turret budge in either direction. It meant that, to stand any chance of hitting a chosen target, his driver would have to instantly change course, align the 60-ton monster on to a precise bearing and then bring the Tiger to a halt for firing, while at the same time affording them the best protection possible from enemy fire.

Only the very best tank drivers in the German army were chosen to operate the Tigers. But even these gifted few were not super-humans. What the commander would be asking of his young driver was a near impossibility. It would place his whole crew – who had time and again proven their bravery and risked their lives in battle under his command – in a most perilous position for the sake of a foolhardy gesture he was certain Major August Seidensticker would never endorse. Luck had been with him so far in this war, but he knew he couldn't count on it holding out indefinitely.

Only a half-hour earlier, another shot had struck the gun's mantlet, the protective armour around its lower barrel, and a fragment of shrapnel had gouged a small furrow across his left temple. There was dried blood on the back of his right hand from where he had rubbed the wound. Looking at it, the commander reflected that just a one-inch difference in the shrapnel's trajectory would have sent it ripping through his brain. That would have been game over.

They had already destroyed one Churchill and put another out of action. He had seen the enemy infantry retreating from the hill they had captured and his own countrymen advancing to retake the position. Many were dead or wounded, but the whole damned day of fighting had ended in stalemate.

The Second Lieutenant made his decision and spoke into his radio mike. 'Climb back aboard. We're going home.'

He grinned as cheers came from the crewmen in the hull below.

Using the metal step and handle unique to the 131, his driver clambered up the Tiger's front, where he had been inspecting the damage. Moments later, he was about to slip in through the open front hatch when he glanced at his commander, smiling down from his standing position in the main cupola.

'Behind you,' he yelled.

The Lieutenant quickly glanced over his shoulder and saw a Churchill tank lumbering straight towards them, its gun pointing directly at him.

The driver dropped through the front hatch, the Lieutenant screamed to him through the radio mike, 'Get us out of here right now. The rest of you bail out.'

At the same moment, Doug Lidderdale opened fire with his Sten gun, and Sergeant Shaw joined in with the Churchill's fixed machine gun.

As the Untersturmführer twisted to face towards Doug, he received a round high in his right shoulder. It must have missed bone and gone straight through, because Doug saw blood and torn flesh explode from the German's back and his mouth snap wide open in an inaudible scream. The force of the bullet spun him around and he folded on to the seat below, leaving the hatch cover open.

Doug felt his heart race. He was almost gasping for air. It was the first time he had ever shot a living creature, and, though he felt shocked, he had absolutely no regrets about what he had done. Not long ago, he had witnessed this same tank commander deliberately destroy a British tank and kill three of its crew. Given the opportunity, Doug knew that he would cheerfully blow the man's brains out.

It was a kill-or-be-killed situation and he had no doubt who he wanted to come out of this alive.

Doug yelled to Bill Rider, 'Get alongside!'

The Churchill veered right to carry out his order, while Doug climbed on to the hull roof. The Tiger still hadn't moved and, as he stood up, swaying to maintain his balance, Doug saw one of the German crew emerge from behind the turret, an MP40 sub-machine gun gripped in both hands, pointing towards him.

Doug's blood froze. 'Shit.'

He tried to raise his own weapon, knowing already that he was far too late and powerless to defend himself, when the man's body began to convulse and disintegrate as it took the brunt of a long burst of fire from Sam Shaw in the Churchill's forward hatch.

Out of immediate danger, Doug took a quick pace forward and leaped the four feet separating him from the Tiger's hull. He landed heavily on the mudguard and fell forward. Still clutching his weapon, the knuckles of his hands rammed hard into the steel roof. Ignoring the pain and firing off a silent prayer that nothing was broken, he scrambled to his feet, crouching a few feet from the turret and the commander's open hatch cover. He noticed another potential source of danger, a smaller hatch further forward on his right. It was open and appeared to have been smashed apart by shellfire.

Reg Whatley adroitly leaped the gap and landed heavily next to Doug on the Tiger's roof. In Reg's hand was a Smith and Wesson .38 revolver, which he suddenly thrust forward, firing slightly wide of Doug's waist.

Doug heard a scream behind him and spun round in time to see a second member of the Tiger's crew clutch at his throat. A fountain of blood forced its way past the man's fingers as he hopelessly tried to seal the hole torn by Reg's round.

The solder had appeared from behind the Tiger's turret, where Doug now assumed there must be another escape hatch. He stared at the dying man, watching the life fade in the German soldier's eyes and his lifeless fingers drop from his throat as he slid, feet first off the hull to the ground.

'Thanks,' he yelled to Reg.

Lieutenant Whatley seemed to be in his element – a real-life Bulldog Drummond. Whatley had hidden depths, thought Doug. So meek and mild on the surface, but so quick and fearless in action.

Kneeling next to him, Reg grinned an acknowledgement.

And thanks from Kate too, thought Doug. Both to Reg and to Sam Shaw. Without their fast reactions, she could have been made a widow twice in the past 30 seconds.

Later, there would be time to dwell on the might-have-beens and to thank his teammates properly for saving his life, but, for the moment, there were more pressing matters on which to concentrate. What had become of the Tiger's remaining crew-members? There was no sign of them or their wounded commander.

Probably in the hull, thought Doug. This meant they must get them out without damaging the Tiger's insides. A grenade dropped through the open hatch would do the trick and kill them all instantly, but that would cause immense damage to the new and secret German tank technology this exercise was intended to capture. Furthermore, a grenade had the potential to detonate some of the Tiger's shells and blow it, and themselves, to Kingdom Come.

Doug was pondering this problem as Rider braked the Churchill and brought it sideways on to the front of the Tiger, under the front portion of its huge 88mm gun.

'Have the blighters gone to earth, sir?' shouted Rider.

'I think so, Corporal. And getting them out isn't going to be easy. It's going to be dark soon and then, I imagine, their friends will come looking for them.'

'We know from experience that they're not going to leave a fully intact Tiger on the battlefield for us to nab,' said Rider. 'They'll want to drag it back behind their lines to safety.'

'Yes, or they may be planning to destroy it,' warned Doug. 'They've done that before.'

Doug's expression changed to one of resolve. 'So, if we're going to get these blokes out of here, we'd better come up with an answer pretty damned sharpish.'

itler's ferocious champion – the Tiger tank was a massive threat to the Allies in a vital
riod of the war. Churchill's stunning solution was to snatch a Tiger and use its secrets
ainst the enemy.

The man chosen to carry out this dangerous assignment was Major Douglas Lidderdale. Gagged, during his lifetime, from talking about one of World War II's greatest untold stories, Doug left a detailed diary of his secret mission with his son, David.

My darling Douglas,

David telephoned last night to tell me you would be having your operations today. I do hope all prays you will soon be well and fit again. I miss you so much and love you so much, if I had been fit and well would I along to see you. David has always called on me and given me your news. I think the Doctors and Nurses have been very kind.

We have now been Married 50 years and still love and adore each other.

All my love to you darling

Kate.

Beautiful and vivacious dancer, Kathleen Crane, was a bride of just two days when Churchill sent Doug off to Tunisia to catch him a Tiger, unsure whether he would ever see his adored Kate again. Their passionate wartime love letters were witness to a devotion which would last more than half a century until her death in 1995.

Top: Lidderdale's beloved handbuilt, two-seater sports car – nicknamed 'The Beast' – remained in England with Kathleen while Doug went in search of a very different animal.

Bottom: The image of his prey, logo of the German Heavy Tank Battalion 501, was affixed to Doug's diaries as a constant reminder of his mission.

ƆTCHA! Less than 24 hours after Doug and his team's deathly struggle with Tiger
1's crew, 21st and 25th Tank Brigade personnel inspect their trophy (*top*). On learning
⊥t Doug had caught his Tiger, Churchill dashed to Tunis to inspect his quarry first
nd. (*Bottom*) He is seen on the Tiger's turret, facing Doug (back to camera) and
⊥ding one of 131's 88mm high explosive shells.

Example of the many orders received by Major Lidderdale. (*Top*) The missive announcing Churchill's arrival to view the tiger, and (*bottom*) the order to move the Tiger to Tunis racecourse workshop to prepare for its move to England.

ove: At great personal risk, King George VI (*middle*) flew to Tunis to see the
orious Tiger at first hand and congratulate Doug and his men.

ow: Heroes. In November 1943, Major Lidderdale, Lieutenant Corporal Pumfrey and
ver Wilkes stand proudly on Horse Guards Parade, London. in front of the captured
er 131.

Above: Researching a great story (*left to right*) Authors Noel Botham and Bruce Montague with Major Lidderdale's son Dave Travis and an archivist from the Boving Tank Museum in Dorset.

Below: Noel, Dave and Bruce stand in front of the Tiger, a tank that has come to mea so much to them.

He frowned and went on grimly, half to himself, 'Maybe we should blow it up ourselves with them still in it. At least there'll be one less damned Tiger for Jerry to use against us.'

Reg Whatley, who had now risen to his feet, said, 'Actually, sir, you're wrong. If you turn round now, you'll see our chums scarpering off, back the way we came.'

Doug spun round and saw the backs of three men hurrying away from them – the two on the outside supporting the one in the middle. He had an arm around each of their shoulders and his feet were dragging.

Doug muttered, 'God damn it! That must be the one I shot.'

He got no further before Sergeant Shaw yelled, 'Shall we let 'em have it, sir?'

Using the Tiger's gun as a handrail, Shaw had crossed from the Churchill and joined Reg and Doug on the Tiger's hull roof. Shaw's Sten was at the ready.

'Damned right, Sergeant,' shouted Doug. 'Shoot the bastards.'

The two Sten guns opened fire simultaneously and Reg began firing his revolver at the men who, with urgent backward glances, plunged sideways into the brush, dragging their wounded commander with them. The three British soldiers continued to fire at the spot where the men had dived for cover until their ammunition was exhausted.

'I think I got one of them, sir, but I can't be certain,' said Shaw, slipping a fresh ammunition clip on to his Sten gun. 'They didn't seem to be carrying any weapons but you never know. If we go out there looking for them, we could come a cropper.'

'I think you're right, Sergeant,' replied Doug. 'It can't be too difficult, with our knowledge of tanks, to work out how to get this brute moving. Let's shift it away from here so even if Jerry does come back he can't get his hands on it. Who wants to try to push the boat out?'

'I don't mind if I do, sir,' said Reg, mimicking Colonel Chinstrap, Jack Train's much-loved character in the popular radio show *ITMA*. 'Let me get down there. The engine's still running so that's half the problem solved. The gears and transmission can't be that different from ours, and the clutch must be pretty obvious. Leave it to me. What a bloody wheeze, eh? To be the first Englishman to drive a Tiger.'

The Lieutenant's enthusiasm was infectious and Doug found himself smiling. Then his face took on a serious look. 'Be damned careful down there, Reggie. We didn't give them much time to get up to mischief, but they might have booby-trapped the thing. Take a few minutes to look around before you touch anything. We'd look very stupid if we got ourselves blown to smithereens by being careless.'

'Right,' said Reg, swinging his legs over the hatch rim, his wide grin still in place. 'But they seemed more interested in bailing out of this beast than in destroying it. Though I'll still be careful. I promise.'

He dropped down to the command pedestal before he slowly lowered his whole body into the turret room.

'I'll be down to join you in a few seconds, Reg,' Doug called down through the hatch.

Turning to Sergeant Shaw, he said, 'Most of our lot have already withdrawn, so let's take our Tiger under darkness as close to our lines as we can – short of the bloody minefields, that is. At the first streak of daylight tomorrow, we can take it back to base when our chaps with the six-pounders won't mistake us for the enemy.

'Tell Corporal Rider to drive the Churchill –' he stopped as he saw Sam Shaw's crestfallen expression '– all right, Bessie! – and lead the way for us. We'll park this big boy near the damaged Churchill. Jerry wouldn't venture that close to our lines even if he knew exactly where it would be.'

'Do you want me to stay with Rider, sir, or come with you and Mr Whatley?'

'Better stay with Rider, and keep close to the machine gun. There could be pockets of enemy troops located along our route. They will probably think you're being chased by one of their Tigers, but we can't be too careful.'

'Sah!' Shaw snapped, and leaped from the Tiger's hull across to the Churchill.

Doug waited until the British tank had reversed and swung forward to idle, pointing away from him, and about 20 yards in front. He raised an open palm to Shaw who was standing in the Churchill's open command hatch, indicating he should wait, and then lowered himself into the Tiger's turret basket.

His first impression was that the basket was much larger than the Churchill equivalent. The pedestal on which he stood was about four feet above the turret platform, and folded up to form a back rest for the commander's lower seat during closed-hatch fighting. To the right of this was a sheet metal guard which he assumed would protect the commander from the 88mm gun's back blast when the breech opened.

Below and slightly in front of him was a second seat, which, from its position, he assumed was for the gunner. Further forward and lower down was Reg Whatley, his lanky frame folded into a cramped seat behind a half steering wheel, his fingers fluttering over a host of knobs and levers.

'Figured it out yet, Reg?' asked Doug.

The Lieutenant looked over his shoulder and grinned. 'Just about, sir. I must say, it's a bit of a tight fit down here. Legs hemmed in and no adjustment on the seat height. I'd hate to be the poor bugger who has to sit here for hours on end. Must end up with so many stiff joints and cramp he probably can't even stand up. There's also no way you can stick your head out and have a clear view of where you're going.'

Doug peered round in the gloom of the driver's compartment. The only view available to the driver was through a glass-covered visor less than a foot wide and about three inches high.

'And about six inches thick,' said Reg. 'It's got a clutch and a gear lever and seems to have about eight or nine gears – though I've no idea how many are forward and how many are reverse. I think I'll just have to push it in the first slot and hope that that's right.'

His grin widened. 'It'll be interesting to find out if this 60-ton beast does kangaroo jumps if I get the clutch and gear synchronisation wrong. Oh, by the way, sir, their radio is smashed to pieces and there's some shrapnel damage above where I'm sitting. It would seem that part of the round which disabled their turret might have deflected down here. There's also quite a bit of blood splashed around. Some of it probably belongs to the commander, but I think one of the crew might have been injured when their turret was done in.'

Doug eased himself past the gunner's seat to peer over Reg's shoulder. 'Let's put aside the charming question of whose blood it is and get this blighter on the move. The only way we're going to find out if you're right is if you try it. But it would be bloody marvellous if you can do it without stalling the engine, old chap. We don't want to have to figure out how to start the damned thing.'

'Right,' said Reg. 'Here goes.'

He gripped the black-knobbed gear lever in his right hand, and was about to slide it into place when a staccato burst of nearby sub-machine gunfire carried through the open commander's hatch behind Doug.

'That doesn't sound too healthy,' he rasped, staying

crouched and edging backwards towards the cupola. He climbed on to the commander's top seat. 'Hold it where you are, Reg, and wait for my signal. I have to take a peek up top and see what's going on.'

He slowly raised his eyes above the cupola's drum-shaped top as there came a further burst of fire, blending almost instantly with the dull metallic zap of rounds striking the inside of the raised hatch cover and the Tiger's protective armour.

'Christ!' he gasped out loud, and instinctively ducked down. 'I almost felt one of those part my bloody hair.'

Almost immediately following the enemy fire came a further lengthy burst from a Sten gun, and the bellowing voice of Sergeant Shaw.

'Sorry about that but I've got them pinned down now, sir. It looks like the bastards who abandoned the Tiger. They're in a gully off to your left. Two of them by the looks of it, so the man you winged has either croaked or he's in a bad way and laying up somewhere. I suggest you climb out while I cover you and slide down the side of the Tiger on the blind side.'

Taking a deep breath, Doug hauled himself out of the rear escape hatch and rolled across the hull roof. Glancing in the direction the Tiger was pointing, he spotted Shaw lying on the Churchill's hull, directing his Sten over the turret towards the brush to the west. Shaw strapped the Sten gun around his neck and gave Doug a go-ahead wave with his left hand. Doug gripped the edge of the mudguard and, using the tank's tracks as a step, lowered himself four feet to the ground.

Removing the Sten from his neck, he moved to the front of the Tiger and peered around the four-inch-thick nose plate of the hull. A sub-machine gun began firing again and Doug could hear the lead impacting on the Churchill's superstructure.

The Germans had approached rapidly along the gully and

were nearly abreast of the British tank, forcing Shaw to kneel to see them over the turret.

Seizing this opportunity, while the enemy fire was concentrated on Shaw, Doug sent a long burst from his own Sten gun at the spot in the brush from where the German rounds were coming. He was rewarded with a shriek of pain and the MP40 fell silent. Clearly believing that Doug presented his biggest threat, the other Panzer crewman opened fire in the Tiger's direction, squeezing off single shots that zapped against the tank's front panel, frighteningly close to his head.

Doug stepped back behind cover just as a hand grenade, lobbed from the Churchill by Shaw, exploded with a resounding blast, instantly followed by a short loud scream. Doug waited a full minute before peering round the Tiger's front. He squeezed the trigger rapidly twice, sending a couple of rounds into the brush, but there was no answering fire.

Ahead, he could see Shaw, crouching on the Churchill's hull, Sten gun at the ready, peering down at the enemy position.

The Sergeant turned and held one hand up, thumb extended. 'I think that's finished them off, sah.'

'OK, Sergeant. I'm going forward to check out the situation. But keep me covered.'

With his Sten gun held in front of him, Doug walked slowly towards the brush along the edge of the ridge, ready to throw himself to the ground at the first sign of movement. The carnage he discovered made his stomach heave and he had to turn his head away. After a couple of deep breaths, he forced himself to look again.

One German was spread-eagled on his back, mouth wide open and eyes staring, his chest soaked in blood where Doug's Sten burst had raked him from hip to opposite shoulder. His body was undamaged by the grenade blast which had devastated his kneeling companion. The man had lost both his legs at the knee and his stomach had been ripped open. He had lived only long enough to utter that brief strangled scream before he died.

Of the commander there was no sign. Either he'd been left behind along the track and was waiting to be rescued, or he was dead. Doug cared little either way. He climbed back on to the Tiger's hull. He called across to Sam Shaw who was waiting by his Bessie.

'Got your canteen with you?'

'Of course, sah. It's only water though.'

'That's all we need for what I have in mind. Mr Whatley, unscrew your water canteen if you please.'

As the Lieutenant complied, Doug did the same with his own. 'Gentlemen, I propose a toast. To the capture of the Tiger – Herr Hitler's favourite toy. A day late perhaps, but what a present for his birthday!'

They all swigged from their canteens. Then Doug shouted out to his Sergeant with renewed vigour, 'Same plan as before. We'll follow you down. Get moving as soon as we are under way and we'll try not to run over you.'

Doug dropped through the commander's hatch. 'Time to try that gear position, Reggie. And let's hope you get it right. I don't want us hanging around here any longer.'

With the clutch depressed, and firmly holding the steering wheel in his left hand, Reg used his free hand to push the gear lever into the first slot before cautiously freeing the clutch. The Tiger shuddered and grated and then slowly began clanking forward.

Reg turned a beaming face towards Doug. 'Wise choice, thank God! I would have hated having to say "oops"!'

Doug beamed back. 'Well done, old chap. Now we need to get

it into second gear. At this rate, it'll take us till tomorrow night to get where we are going.'

Selecting the second notch and easing off the clutch slowly, Reg felt the Tiger pick up the pace. 'Doesn't seem to even need the clutch now it's going. I wonder how many gears this bloody contraption has. There are an awful lot of notches.'

Reg took it up to fifth before Doug decided they were progressing at about 10mph.

'That'll do nicely, Reg. It's going to be dark pretty soon and we don't want to run full tilt into something nasty at this stage. Having caught our Tiger, I don't want to let it go. We've almost managed to pull off the impossible. Think how bloody stupid we'd feel if we ballsed it up now.'

Familiarising himself with the controls, Reg was feeling far from stupid. This was pure exhilaration. He had fought like the tank's namesake to win their prize. In the process he had killed, for the first time, a young man of his own age. And now he had mastered this great, clanking monster's controls. He had tamed the beast. He beamed over his shoulder towards Doug.

'Whatever you say, sir. I know it's probably wrong after all we've been through, but I feel marvellously bloody happy.' He laughed out loud. 'This makes foxhunting seem very bloody tame in comparison.'

Doug found himself laughing too, but soon groaned as pain lanced through his right knee. The adrenaline had stopped running and his leg was throbbing from where he had knocked it jumping aboard the Tiger. He rolled up his trouser leg to examine the damage. It wasn't pretty, but it wasn't so bad. Doug paused and grinned again.

At least I'm not dead, he thought. After what we've just gone through, that has to be a bonus.

CHAPTER EIGHTEEN
21 APRIL 1943 –
1800 HOURS

The two tanks clanked their way towards the British lines as the sun finally disappeared over the western mountains. As was usual in these latitudes, darkness fell almost immediately.

Doug called a halt near the crest of a slope next to an abandoned foxhole, flashing a torch signal to Sergeant Shaw in the Churchill's command module ahead. It was already turning cold and he felt dampness in the air.

'I wouldn't be surprised if it didn't start raining again,' he told Reg as the Lieutenant clambered through the hatch and joined him on the Tiger's roof. 'I'm going to leave Shaw here overnight. I don't think there's any chance of Jerry coming this far over but the bloody Arabs are another thing altogether. By morning, they could have stripped it of everything that can be cut away. Meanwhile, I need to talk to you all. Let's get the others over here. Have you worked out how to turn this thing off?'

'Pretty simple, actually.' Reg laughed. 'There's a switch on the side. Right next to a button, which is what I presume you press to start it.'

A few minutes later, Shaw and Rider appeared out of the darkness. Doug and Reg dropped to the ground to join them.

'Well, sah, old Yum Yum is going to be pleased,' said Shaw.

'Old Yum Yum?'

'Yes, sah. That's what my old dad calls Mr Churchill, sah. Don't know why, but a lot of the old 'uns call him that.'

'Well, I've heard him called worse,' said Doug, grinning. 'And yes, I think he's going to be very pleased indeed. But we need to concentrate on now.

'Every one of you has done incredibly well today. You all know how close we came to cashing in, and I have to thank you, Shaw, and you, Reg, for saving my life. It was a damned close call, and not one I'd like to repeat.'

Shaw coloured and shuffled his feet and Reg gripped Doug's arm. 'If you hadn't taken out the Jerry commander and that chap in the ditch, then we might all have got it, sir. Thank God, it turned out the way it did.'

'Good of you to say so, Reg, but thank you both anyway. Which means that what I'm about to ask you now is a bit hard to take.'

The three men looked at him expectantly.

'Until I've discussed it with our bosses, I don't want any of you to breathe a word about what has gone on today. Nobody here, apart from the Brigadier, knows about our mission for the Prime Minister, and without his OK nobody must know. Not ever. Understood?'

His three listeners nodded their heads.

'I'm going straight from here to talk with your CO, Brigadier Maxwell, at the 25th. It won't stay a secret forever, and the right man will know exactly what part you played in all this. But for the moment we are engineers, not these new bloody commando types. Derring-do is not what we're about. Once I get the word from above, then you're all free to talk about it. Until I've done that none of you must breathe a word to a soul. Is that clear?'

The men nodded again and Shaw let out an energetic 'Sah!'

'OK then. Sergeant, I want you to stay here and keep an eye on our catch. I'll be back in the morning. But, if anyone else comes along before I get back, then you just happened to stumble on our Tiger.

'Now, Corporal Rider, you and me and the Lieutenant here are going to take old Bessie back to the 104 and I'm going to have a chat with Brigadier Maxwell. But not before we have a well-earned mug of tea. Now let's make sure we haven't left any tell-tale bits and pieces in the Tiger and we'll be on our way.'

CHAPTER NINETEEN
21 APRIL 1943 – MIDNIGHT

Brigadier Richard Maxwell, commander of the 25th Tank Brigade, was seated behind a makeshift desk in head-quarters command tent when Doug was shown in by one of his aides-de-camp.

Maxwell, a tall, athletically built man with a hint of grey at his temples, failed to raise any hint of a welcoming smile as he waved Doug to a seat.

'Probably no need to tell you that things haven't gone too well for us today.' He spoke with a hint of Irish burr in his soft voice. 'I'm told the 48th had four tanks knocked out and at least three of their chaps killed. I just pray that young Lott pulls through but it doesn't look good.

'On top of that, the East Surrey's CO, Hugh Bruno, was killed and several other officers wounded. Not much to cheer about on their first day in action. Let's just hope we've learned something from today's mistakes. They're pulling back five or six miles to Sloughia on the Testour road, though I suspect they're going to be called on again pretty soon. Right now, it's the turn of the North Irish Horse, who are about to make a big push on Longstop.

'I guess, though, that we're keeping your chaps busy. We keep breaking them and your lot keep putting them back together.'

'Yes, sir,' said Doug, 'but that isn't what I came to see you about. I have some good news for you, for a change, and I also have a bit of a problem.'

'Go ahead, let's hear it.'

'As you're aware, sir, I was given certain specific instructions when I was sent out here.'

Brigadier Maxwell leaned forward, tired eyes instantly more alert. 'Yes, I remember you telling me about that.'

'Well, it's become something of a long story. But here goes.'

For the next five minutes, Doug took the Brigadier through his briefing in London and his unit's sudden departure for Tunisia. Maxwell remained silent until Doug reached the part about capturing the Tiger that afternoon.

The look on his face was one of incredulity. 'You mean to say that you and your REME chaps actually captured a Tiger tank intact?'

'Well, one of them, Lieutenant Whatley, you will remember, is with your headquarters workshop.'

'Yes, but you managed to overpower the crew in the middle of today's scrap with none of you being hurt.'

Doug nodded. 'Yes, sir, we did.'

'Miraculous! Where is it now?'

'We brought it back to the edge of our lines, sir. I've left my Sergeant there to guard it.'

'You think that Jerry will try to steal it back?'

'I wasn't thinking so much about the Germans, sir. I don't think they would risk coming in that far. But the locals – they might take it into their heads to use it as scrap metal.'

It was Maxwell's turn to nod. 'Quite right, Major. Well, what do you want from me?'

'Your advice, really. I've told my fellows not to say anything at all about our action today. It's not the usual kind of thing for engineers to get involved in a shooting match with the enemy. I'm worried that it might not go down too well in some quarters.'

'You're probably right, Major. A shrewd observation. All the same, nobody is going to question your courage. Ha! Damned brave, if you ask me. But discretion before valour, eh? I'd better take advice on this from the Major General. Come and see me first thing in the morning. Hopefully, I'll have an answer for you by then. And we'd better get someone out there to relieve your Sergeant, don't you think?'

Doug did not seem enthusiastic. 'If we can keep this amongst ourselves for the time being, sir.'

'It's going to come out sooner or later. We can't hide a 50-ton tank forever, you know.'

'It's closer to 60, sir.'

'Good God!'

Maxwell stood and grasped Doug's hand. 'Well done, Major. At least you've managed to cheer me up. It's the best news I've heard all day.'

The next morning, after an early breakfast, Doug took a GPW Jeep across to Brigade headquarters where he found several high-ranking officers gathered near the command tent.

Brigadier Maxwell broke away from the group and came over as Doug climbed from the Jeep and saluted. He returned the salute and told Doug to walk with him to the edge of the camp.

'The top brass is showing a hell of a lot of interest in your Tiger, Lidderdale. Seems someone in HQ does know what you're doing here. It's not only the technical know-how they're interested in – it is the propaganda side as well. Big boost to morale, they say.'

'I hadn't quite thought of it in that way, sir.'

'Oh, yes. No doubt about it at all. You've pulled off quite a coup.'

'Thank you, sir.'

'Don't thank me. I'm just the messenger. Even though you're technically with the 25th and this was more of a 21st action, they want you to take over full responsibility for the Tiger. In fact, you are to carry out your original orders and take it back to England.'

'Do I get a boat, sir?'

'I suppose so. This is a unique situation, you know. We need to rewrite the rulebook.' Maxwell looked like a junior officer again – years of crust seemed to have fallen off him. 'Oh, and another thing. They are already talking about a propaganda film being shot here, and I expect the top brass will want to see it. Even General Alexander, and Eisenhower himself.'

'That's encouraging, sir.'

'But there's one thing your chaps are probably not going to like. The powers that be don't think it will sit right with the public if we say the crew was overpowered by a group of engineers.'

'But it was, sir.'

'I know it was, Lidderdale. But it wouldn't look so good that, in a battle involving infantry, artillery and our tanks, the most successful victory was achieved by non-combatants.'

'We're trained to fight, sir.'

'Yes, yes, that goes without saying. But we don't consider you front-line troops. Whatley's specialisation, for example, is in tank recovery. Yours is engines and repairs. You say the Tiger's turret was damaged. Well, that is to be the story. When it became inoperable because of a well-aimed shot from one of our Churchills, we must assume the crew decided to abandon it.

'Your real report can go to the REME Director in London but I'm afraid it's probably going to remain secret until after the war is over. That's the decision from the top. Can you trust your men not to talk?'

'Of course I can, sir. And I'll make damned sure they get all the credit in my report. It just seems a shame people won't know what these chaps have actually achieved until all this is over.'

'The people who matter will know, Lidderdale. At the moment, that's all that counts. Meanwhile, you'd better relieve that Sergeant of yours and leave it to our chaps to discover this Tiger. Abandoned of course.'

'As you wish, sir.'

'It may also be that you can't recover it for a few more days because of the big push going on at the moment. But rest assured, Major, it's your Tiger and nobody else is going to get their hands on it. I'll make certain you receive orders to that effect.'

Later that morning, Doug and Sergeant Shaw were silent observers as members of the 48th Battalion and the 21st Brigade discovered the intact Tiger which had appeared in front of their lines as though by magic. A few German shells landed in the area but it appeared that the enemy, who had fought so fiercely to defend the hills to the east, were now withdrawing – for reasons best known to themselves – from the ground which had seen the shedding of so much blood on both sides in the past 24 hours.

No one was able to explain how the Tiger came to be

abandoned so close to the British lines, but the 48th were claiming, quite rightly, that it was a shot from one of their tanks which had incapacitated the mighty beast and forced its crew to jump ship.

'Excuse me saying so, sir,' said Shaw, glaring disdainfully at the dozen or so officers and men clambering over the subdued Tiger, 'but what they are saying sounds like a load of old bollocks, sir.'

'Couldn't agree with you more, Sergeant, but that is the official line of bollocks you're hearing, and that's the only line to take at the moment.'

Shaw exhaled noisily.

Doug went on. 'Let's leave them all to it. We'll come back and retrieve our Tiger another day. Meanwhile, again, I think a nice mug of tea might be just the ticket.'

CHAPTER TWENTY

6 MAY 1943

General Alexander had promised Churchill a speedy end to the North African campaign and he was determined to fulfil it. That way, he could put all his concentration into the real war in Europe and the Pacific.

With 250,000 Axis troops bottled into the northeastern quadrant of Tunisia, Alexander refused to concede a single day's delay in the launch of Operation Vulcan.

For Doug, this resulted in his not seeing his Tiger again for two frustrating weeks. He was tied up either in the field of battle or in the 104's field workshop and casualty park at what the Allies had nicknamed 'Scorpion Corner' in Redcap Alley, near Medjez.

Alexander dismissed the German offensive of the previous two days as 'an irritating sideshow', an impetuous and desperate folly that had cost von Arnim dearly in men and equipment, including the loss of all the 10th Panzer Division's tanks.

But, if Alexander expected the Axis forces to crumble under the concerted onslaught of British, American and French armies, he had badly miscalculated. After a continuous 48-hour attack, the Germans, using their consummate skill in improvised defence, held their positions and inflicted serious losses on the Allies.

Some of the heaviest casualties were brought about by Tiger tanks. Allied tank losses were in their hundreds during the first week of Operation Vulcan. This final 'mopping up' phase of the war in North Africa saw some of the fiercest fighting of the desert campaign with a huge surge in the numbers killed or injured on both sides.

In the thick of it were the field units of the REME and RACE, responsible for rescuing damaged tanks and their crews from the worst storm centres of the battlefield. Among them were Doug Lidderdale and Reg Whatley and their fellow Tiger hunters, AQMS Shaw and Corporal Rider, doing what Doug now described as 'our day job'. They constantly risked their lives to retrieve and repair the battered mechanised weapons on which the Tank Brigade depended to crush the enemy.

By 6 May, General Alexander, growing increasingly impatient and frustrated, decided to change tactics. In the south, Montgomery's 8th Army would maintain diversionary local action. In the north, the US II Corps would tie up the Axis forces and keep them pinned down. But surrounding Tunis, in the centre, the 1st Army's IX Corps, under its new commander General Brian Horrocks, would launch an all-out attack, with mass air and artillery support.

Operation Strike started before dawn, east of the key town of Medjez-el-Bab, some 30 miles from Tunis. It involved the bulk of Horrocks' force, including four tank contingents.

By midday, the Allies had punched a massive hole through the German defences. Despite being under constant artillery bombardment, only a handful of enemy tanks were sighted.

The Germans Panzers were playing a defensive game. As soon as they were under heavy attack, they retreated.

By nightfall, Horrocks' men were within sight of El Bathan, less than 15 miles from the outskirts of Tunis.

Though successful, the British thrust had not been achieved cheaply. The sheer weight and accuracy of German artillery fire had cost the British dearly in tanks and lives, and resulted in many individual acts of bravery.

Lieutenant Reginald Whatley was one such man to be singled out by his commanders for an award. He personally led his team on to the battlefield on 12 separate occasions to rescue Churchills that had been immobilised or set ablaze by enemy rounds. Most of the Churchills were in the midst of minefields. To add to the danger, Whatley's recovery vehicle drew constant enemy shell and mortar fire throughout the rescue operations.

The unflinching courageousness of his actions and total disregard for his own safety meant the difference between life and death for more than a score of his comrades-in-arms who crewed the stricken tanks. For this, he would receive the Military Cross, an honour reserved for junior officers for gallantry in action.

On the following day – 7 May – Winston Churchill was able to summon his top aides in London to share a magnum of his favourite Pol Roger champagne and announce a double victory in Tunisia. General Horrocks had continued his advance on Tunis and captured the city virtually unopposed. At almost the same moment, the American II Corps had more than made up for their defeat at Kasserine by taking Bizerta, Tunisia's second most important port.

The German supply route from Italy was effectively severed. 'The rope is now around the Afrikakorps' collective neck,' a chuckling Prime Minister told his aides as they chinked glasses. 'Von Arnim may not realise it yet, but the trap door beneath is already opening.'

Unaware of the historic events taking place on the north coast, Doug Lidderdale had his own reason to celebrate. After the Allied breakthrough on 6 May, word had come from 25th Brigade headquarters that he could now be spared to concentrate all his energies on getting his Tiger back to England. Its first appearance would be as the star of a War Ministry film that would be rushed back to London to be viewed by the Prime Minister and his cabinet. Not just a few snapshots. They wanted to see the Tiger in action.

CHAPTER TWENTY-ONE

MAY 1943

To Doug's utter delight, Reg Whatley had responded to a radio message and turned up at the 104 shortly before midnight after a heroic and exhausting day with Horrocks' advancing army.

'Showing off the Tiger wouldn't be the same without you,' said Doug, slapping Reg on the shoulder, his mouth widening in a huge welcoming grin.

Reg returned the grin and vigorously shook Doug's hand in both of his. 'You didn't think a little thing like Hitler's war hounds were going to stop me sharing the fun part of all this, did you?'

'Never doubted it, old chap. Shaw and Rider said they were prepared to put their wages on you being here.'

After only five hours' sleep at the REME base camp, Reg had risen with the rest of Doug's crew and driven through the early-morning rain in a scout car to where he had parked Tiger 131 a fortnight before.

They were accompanied by a giant D8 – a 140hp caterpillar

tractor that was hooked up to the front of the Tiger by twin tow bars.

Reg and Sergeant Shaw – both still officially attached to the 25th Tank Brigade's REME workshop – elected to ride on the Tiger's turret. Doug and Corporal Rider were in the scout car leading the 36-ton tractor, which was dwarfed by the massive German tank hitched to its rear end.

Once they had their Tiger safely harboured in the 104's field workshop at Scorpion Corner, Doug and his men spent a full week getting to know and master their captive's unique capabilities. They also repaired the turret's turning mechanism and replaced the damaged forward hatch with one from a gutted Tiger found on another battlefield.

By 13 May, Tiger 131 was in full working order, its fuel tank filled, its various oil chambers topped to the line, and ready to make its debut as the villainous star in its very own movie. At 1600 hours, Doug gathered the whole of 104's contingent together to personally thank them for a job well done.

At the moment he was about to speak, a priority message was handed to Doug: General Harold Alexander, commander of the 1st and 8th Armies, had just signalled Churchill to report that the Tunisian campaign was finally over. All enemy resistance had ceased and the Allies were now masters of the North African shores.

For the first time in almost three years, the guns in North Africa had fallen silent. In all, more than 70,000 Axis troops and 65,000 Allies had been killed or wounded in the North African campaign, and its end meant that for more than 200,000 surrendered Axis troops the war was over. For the Allies, however, the fight to take back Europe was about to begin, with the defeat of the Japanese being another significant problem ahead of them.

When Doug read out the message, his men threw their berets in the air and cheered until their throats were sore, slapping each other on the back and waving their right hands, fingers extended in Churchill's famous V sign.

When the men finally fell silent, Doug read out the second section of the message, which was almost equally amazing: General Alexander's top priority, after accepting the German surrender, was not to begin tackling the huge administrative nightmare of confining and feeding 200,000 prisoners, nor to concentrate on the rumoured invasion of Italy. Instead, he was to visit just one celebrated German prisoner of war – the captive Tiger.

'He's coming here tomorrow to watch us put our pet cat through its paces,' Doug announced. 'And he's bringing General Ron Weeks, Deputy Chief of the Imperial General Staff, with him.'

The film, shot on 16mm film in Redcap Alley, required the Tiger to be put through a demanding series of exercises, with Doug himself doing most of the driving. The army's film director wanted it to appear menacingly over a hilltop, crashing its way through a giant haystack and running with its snorkel exhaust – a device for supplying air to the engine and crew when the tank is fully submerged – in place.

Between shots, the two generals crawled all over the Tiger, excitedly exploring its interior and swapping impressions like a couple of schoolboys. At the end of their visit, General Alexander told Doug he had had more fun than on any other day since Operation Torch began.

'I don't think the PM will be able to resist coming and seeing this brute first hand,' he said, smiling broadly. 'And I wouldn't put it past him wanting to fire the big gun either. It's a new toy and, just like the rest of us, he'll want to play with it.'

'Well, sir,' said Doug, 'it's his to do with as he pleases. We'd better make sure it's pointing in the right direction so he can't do any damage.'

On 22 May, Doug's orders to take the Tiger back to England were confirmed from the Deputy Director Mechanical Engineering:

> To: O.C. 104 Tk. Wkshops DADME Tunis Area
> From: DDME First Army Originator's machine:
> ME/998 Date: 22
>
> ARRANGE TO MOVE TIGER TO 12 PORT
> WKSHOP DET RACECOURSE TUNIS KO 256.
> NOTIFY
> THIS HQ TIME OF ARRIVAL. NOMINATE
> ONE OFFICER OR O.R. EXCL YOURSELF
> CAPABLE OF DRIVING TANK TO GO
> TO U.K. IN CHARGE. FURTHER INSTRUCTIONS
> AS TO SHIPMNT WILL BE SENT TO DADME
> TUNIS AREA

Alexander had been quite right. Churchill did want to see his Tiger, not to fire the gun, but to drive it himself. That, and to hear first-hand from Doug just how his Tiger had been captured. The explanation Doug was only too willing to provide, but, even if Churchill was able to squeeze into the cramped driving seat of the Tiger, Doug would not have allowed the Prime Minister to take the Tiger for a spin.

For he believed the massive 12-cylinder 650bhp engine was in danger of seizing up.

The problem had begun on 24 May, nearing the end of the Tiger's 35-mile drive to its temporary home in No. 12 Port workshop, close to the Sea of Tunis on the outskirts of the newly liberated Tunisian capital.

It was a bizarre journey along roads that had witnessed ferocious fighting just two short weeks ago. The evidence of these desperate battles – the devastated villages, burned-out and twisted trucks and tanks – was everywhere. Incongruously, it seemed to Doug, who viewed the detritus of war from his vantage point in the command cupola, there were Arabs working in the fields among the wreckage as though nothing of import had ever happened here.

His deliberations over the odd after-effects of war were interrupted by a cry from Shaw who was taking a spell at the wheel. Doug dropped to the deck below and his eyes followed the direction of Shaw's pointing finger. A thin jet of steam was escaping from the engine compartment.

Doug made a quick decision. 'It doesn't look too bad right now and we're very unlikely to find a tow out here. So we'll keep going. It's only about another couple of miles. Ease off on the accelerator as much as you can. But if it gets any worse we'll have to stop.'

Half an hour later, they limped into their Tunis base workshop where Doug began stripping down the engine. By day's end, he had discovered that a vital rubber sealing ring on the cylinder block had failed. He surmised this was due to overheating which in turn was caused by extended running periods of the engine. It was probably an original design fault.

The solution was either to create a new one himself in the workshop, or to scavenge the part from one of the Tiger graveyards scattered around the northern sector of Tunisia. The Americans had seized the main Tiger repair centre and told Doug bluntly that they were keeping all the spare parts for themselves in case they could get their hands on a Tiger of their own.

Even after he did locate a replacement for the damaged sealing ring, Doug was reluctant to run the engine before he and his men could completely disassemble both it and the gearbox to check the status of all the moving parts. He couldn't run the risk of the engine seizing. Unless the tank could be kept in perfect running order, it would be of zero use to them in assessing, extracting and properly diagnosing the revolutionary technical data contained within the Tiger.

It would also mean that they had risked their lives for nothing, and Doug wasn't about to let that happen.

On 31 May, he received the following order:

TO: MAJOR LIDDERDALE
FROM: LT. COL DAVIS.   Date : 31
CHURCHILL ARRIVING HERE TOMORROW.
WILL
WANT TO SEE TIGER. PROBABLY DRIVE IN IT.
REDOUBT
TO RACECOURSE EARLIEST. IF REQUIRED
WE MAY GET DRIVER TO BRING
IT HERE BUT THIS NOT DECIDED
YET.

Churchill and the Foreign Secretary Anthony Eden were flying in to Grombalia airstrip, south of the city, on the 1st. The Prime Minister would visit his Tiger the following day. Doug replied that this would not be possible. Driving the tank is not an option, he told the local Colonel in charge of planning the visit.

'I'll leave you to tell him that yourself,' came the reply.

Doug was nervous on the day but had already decided to stick to his guns. But one look at the great man when he arrived completely did away with his anxiety. He had forgotten just how well padded the Prime Minister was.

The temperature gauge in the open was already over 100 degrees when Churchill arrived. He was wearing a pith helmet, dark glasses and a tropical suit but still looked hot. He produced a huge pocket-handkerchief with which he continually wiped his neck. His entourage included Mr Eden, together with Generals Alexander and Anderson.

Churchill spotted Doug and marched towards him, accompanied by his foremost military adviser, General Alan Brooke, Chief of the Imperial General Staff. Doug snapped off a smart salute but Churchill waved it down and proffered his hand. 'I'd rather shake your hand, Mr Lidderdale. I think you've done me rather proud.'

Then, turning to General Brooke, he added, 'Brookie, I want you to hear this. So come with us.'

He walked away from the group with his hand on Doug's arm. They stopped in the shadow of a large date palm, and he came straight to the point.

'Well, Mr Lidderdale, tell me what happened.'

'We got it, sir.'

'I can see that, Mr Lidderdale. That's why I'm here. But how did you get it?'

'There was a bit of a fight, sir'

'Ah, I thought so.' He paused to light a large cigar. 'Well, come on, man, don't keep us in suspense. Tell me more.'

'We were forced to cope with the crew.'

'Cope? How?'

'I'm afraid we had to kill them. At least four of them. I think the commander may well have got away, which means they must know we have the tank, sir.'

Churchill's eyes lit up at the mention of a fight. 'Don't be

146

afraid, Mr Lidderdale. They were the enemy. You did your duty. That's what happens in war, as we know all too well. People get killed. Remember what I told you in London. Butcher and bolt. That's how it's done.'

'I do recognise that, sir. I'd like you to know that my men showed extreme bravery during the Tiger's capture. I think this should be recognised. I know we can't talk about it publicly now, but if it hadn't been for them I would be dead.'

'One day, Mr Lidderdale, the nation will know all about their courage and yours. But you must not talk about this mission until I give you the word.'

'Of course, sir.'

'This fearless action by you and your men could alter the face of the war. I'm sure the General here will join me in saying you have deservedly earned the respect and admiration of us all. The secrets you unlock from this monstrous machine of war may well help us achieve the final destruction of Hitler's damnable tyranny. When the facts of your deeds are known, you will have the lasting gratitude of all His Majesty's people. In fact – and this must go no further – in a couple of weeks, my boy, I understand you will be hearing that from the King himself.'

'I don't think we can get the Tiger back to London by then, sir,' Doug stammered.

Churchill's heavy jowls widened in a broad grin, and he pointed his cigar at Doug. 'I am aware of that, Mr Lidderdale. But you won't have to go to him. The point is, it is he who is coming to visit you – and your splendid Tiger, of course.'

Doug's mouth dropped open. 'Good God. The King! Here?'

'Don't worry, Mr Lidderdale, he won't bite you. Just keep it to yourself, there's a good chap. I'd be in hot water with our intelligence people if they knew I'd even mentioned it. Wouldn't I, Brookie?'

The Chief of the Imperial General Staff raised his eyebrows and stroked his moustache, looking fondly at his commander and close friend. 'I would venture to say, sir, that the water around you seems to be permanently at boiling point on that score.'

The great man chuckled. 'Wouldn't have it any other way. Let's go and take a closer look at our Tiger, Lidderdale. The General here is usually more interested in looking at birds. He's a keen ornithologist. But this is one beast he is going to find just as fascinating as his feathered friends. Eh, Brookie?'

Doug led the two men over to the tank, where Reg Whatley was standing to attention.

'Sir, General, may I present Lieutenant Whatley who showed exceptional initiative and courage during the capture of the Tiger?'

Reg stepped forward and saluted.

Churchill observed the Lieutenant for a moment, through a cloud of cigar smoke. 'It's always an honour to meet one of our front-line heroes, Mr Whatley. I once made my living writing about brave chaps like you. A long time ago.'

Winston Churchill gazed at the Tiger for a moment. The tank dwarfed them all. 'Thank you for helping Mr Lidderdale get me my Tiger.' He turned to Doug again. 'And thank the rest of your men too. An exceptional result, my boy. Now let's get up on top so I can get a closer look.'

With a ladder, Doug and Reg's assistance, and a lot of puffing, Churchill finally made it to the top of the turret and looked around.

'It looks even bigger from up here. Particularly that gun. What a monster. I'd very much like to see one of the rounds that this thing fires.'

At a nod from Doug, Reg slipped through the commander's

hatch and emerged a few moments later with one of the armour-piercing rounds. He passed it to Doug who gently placed it across Churchill's outstretched arms.

He took the shell and grimaced. 'Heavier than I imagined.'

'It weighs a little over 22 pounds,' Doug told him, standing close so he could take it back if Churchill looked like dropping it.

'So these are the bloody things that wrecked so many of our tanks.'

Yes, sir. These and the high-explosive shells.'

'How many can they carry?'

'There are racks for over 90 of them. A mix of both types.'

'That's a lot of potential damage, Mr Lidderdale.'

'Yes, sir.'

The Prime Minister handed back the shell and a relieved Doug passed it along to Reg Whatley.

'What about below? Can I take a look?'

Doug stared at the Prime Minister's more than ample waistline. 'That might present something of a problem, sir. The command cupola is the widest hatch and that is only 20 inches across.'

'Mmm,' Churchill murmured. 'Must be pretty lean chaps these tank aces. Perhaps I'll have to be content with looking down from the top.'

'Yes, sir. You should get an idea of its size and you'll be able to see the gunner's and the driver's stations from up here.'

Churchill knelt by the side of the open cupola and peered in as Doug pointed out the various visible features. After a few minutes, and with Doug's help, the 68-year-old premier was back on his feet.

'Quite a vehicle, Mr Lidderdale. By the time you get back, I will expect a full report on how it all works.'

'Yes, sir. Of course.'

'Once more, well done. And you too, Mr Whatley. You and the rest of your team have done extremely well. Very good show indeed.'

Five minutes later, Churchill's cavalcade was sweeping out of camp on its way to Grombalia airport where his plane was fuelled and waiting to whisk him back to London. His flying visit to Tunisia had lasted just 36 hours.

Sixteen days later, the King's visit was almost a repeat of Churchill's. The difference was that, by then, Doug had had the opportunity to carry out a further, detailed inspection of the engine, and decided that this time, and for such a special visitor, Tiger 131 could keep its rendezvous with King George under its own steam.

Doug drove the tank himself to the wide palm-tree-lined Avenue Gambetta where the King and Sir James Grigg, the Under-Secretary of State for War, would inspect it. They turned up with Generals Alexander and Anderson once more in tow.

King George VI was as eager as Churchill had been to see the Tiger at close quarters, but needed no assistance climbing the steps on to the Tiger's hull roof. After the introductions were over, King George told Doug, 'The Prime Minister tells me we owe you a big debt of gratitude, Major Lidderdale.'

Doug was aware that, in order to come to Tunisia, the King had made the longest air journey of any British monarch. He had inspected the British 1st and 8th Armies with Montgomery and the other Generals, but he told Doug that this was the moment he has been waiting for.

'I've heard a lot about these Tigers and the damage they have inflicted,' said the King. 'Now, thanks to you, we can find out what makes them so much more dangerous than our own. Is it true, Major, that these Tigers can crush one of our tanks from a distance of more than a mile?'

Doug's reply was cut off by Brigadier Cook from the 1st Army's headquarters: 'From that distance, the Tigers can piss through any of our tanks with ease.'

'So I've heard,' King George smiled. 'But never so graphically put. Thank you, Brigadier.'

Then, turning again to Doug, he said, 'I want you to extend our thanks to your men too, Major. Mr Churchill told me of their courage in capturing this tank. Now that I've seen and learned more about it, I realise just how hard it must have been. You must be very proud of them.'

For Doug, the rest of the royal visit passed in a haze. Wait until I tell Kate about this, he mused. She will be so thrilled. First the Prime Minister and now the King, both thanking me personally. What an amazing day. She's going to be so proud when I tell her.

CHAPTER TWENTY-TWO
JUNE/AUGUST 1943

Shortly after his meeting with the King, Doug received the news they had all been waiting for. The first leg of their transport back to England would pick them up from port La Goullette on Tunis's massive outer harbour wall.

Accompanied by AQMS Shaw and Corporal Rider, Doug drove the Tiger to the port on 29 June, believing the promised tank landing craft would pick them up within days. Stacked on the docks next to the shrouded Tiger were the armour plate, spare bogie wheels and tracks, and 11 crates of spare parts that Doug's team had salvaged in their scavenger raids in May. Enough, Doug hoped, to keep the Tiger running for years to come.

But the days turned into weeks, as all the available tank landing craft were committed to ferrying tanks and troops to Sicily, where the Allied invasion, Operation Husky, had begun on 9 July. Doug's dream of an early reunion with Kate began to fade.

Though he fretted over the long delay, Doug fully understood that the invasion of what Churchill had described as 'the soft underbelly of the Axis forces' must take priority. The task of probing the Tiger's myriad technical secrets was a great antidote against boredom but was not able to lift the despondency that came from the absence of the man who had become his closest friend in Africa.

Reg Whatley, who had been temporarily based at the 25th Tank Brigade headquarters in the nearby coastal town of Hammam-Lif, had been ordered to Algeria to prepare a new location for his unit. The Lieutenant, who gleefully told Doug how much he was anticipating taking part in the much-rumoured invasion of Italy, was dispatched three days before the move, cross-country, out of Tunisia.

Both men found the moment of parting an unexpectedly emotional wrench. In just five months, their friendship had been forged during the innumerable occasions they had faced death together. They had jointly witnessed the extreme horrors of the battlefield and pulled off one of the most daring coups of the North Africa campaign.

As he shook Reg warmly by the hand, Doug found his words were choking in his throat. But in the end they were unnecessary. They didn't need to vocalise the admiration and respect they had for each other.

Finally, Doug said a simple 'Good luck'.

Reg grinned back. 'Same to you, old chap. Let's try to keep in touch.' And he was gone.

Everything seemed to be in flux. Tens of thousands of men and their machines on the move. Doug's own command, the 104 Army Tank workshop, had already been disbanded and most of his men posted to other REME outfits. Even Sam Shaw and Corporal Rider had been reassigned to other units. This meant further difficult farewells with no guarantee of a future reunion. As with Reg, the best Doug could do was wish them good luck.

Doug had been allowed to retain three members of the 104 to assist in getting the Tiger back to England. A Lance Corporal, a craftsman mechanic and a driver.

At last it was their turn to move. The 250-ton landing craft, the LCT 6, lowered its hinged ramp and they were able to drive the Tiger straight on to the main deck of the vessel, which made even the mighty German tank seem small.

A sailor told Doug they could carry a staggering five 30-ton tanks in a single load. At the insistence of Winston Churchill, it had been built to do just that in 1940, and proved once more what a brilliant visionary Britain's wartime Prime Minister was.

With a top speed of 8mph, it took 20 hours to round Cap Zebib and enter the most northerly port of Bizerte. Doug spent much of the time on deck instead of the rather primitive crew quarters, and only visited the mess room for meals. They had clung close to the coast – as U-boats were known to be operating in the straits between Tunisia and Sardinia – and had seen no other ships during their short voyage.

Twice, Doug spotted distant aircraft, too far away to identify as friend or enemy. Aside from the constant throbbing of the three diesel-driven propellers, it was an unremarkable and peaceful journey.

They arrived late at night in the dock area and Doug waited until dawn the next day to unload the Tiger and its spares on to the quay. He noted that many of the nearby buildings had been badly damaged by shellfire and bombing in attacks from both sides as occupation of the city had changed hands.

In the harbour, a large freighter rested on the sea bottom on her starboard side and the mangled superstructure of another, smaller, coastal vessel rose above the water.

Much of the town was cut off from the sea by the large and

ancient fortress wall of a double casbah which gave way to greyish-white and pale-blue flat-roofed houses with wooden jetties where fishing boats and trading dhows were moored. Narrow streets snaked up towards the city centre. Ignoring the war damage, from a distance, the view had a definite eastern charm and beauty.

But Doug knew that up close the aftermath of a brutal and onerous occupation would reveal poverty, greed, a thriving juvenile sex trade and black markets mixed with the inevitable flies and the unique smells of North Africa – the sweet smell of failure, as Doug's friend in Algiers, Ron Parker, would probably have described it.

With the Tiger safely under a tarpaulin, he braved the crowds on foot and walked to Bizerte military headquarters, where the Deputy Director of Mechanical Engineering was able to cheerfully advise him that his time in this accursed city would be mercifully brief.

A Ministry of War transport ship, the *Empire Candida*, was due to arrive in two days. Doug could load his Tiger on 7 August and be on his way to the next port of call on his journey home, to Bone in Algeria.

CHAPTER TWENTY-THREE

AUGUST 1943

Every day brought Douglas Lidderdale closer to England, and every day he grew happier. Doug's newfound exuberance would have been severely dampened, however, had he witnessed the dramatic events unfolding just 150 miles to the west – the precise direction he and the *Empire Candida* were heading.

On board the 6,000-ton British cargo steamer SS *Contractor*, Captain Andrew Brims was in equally high spirits. Convoy GTX-5, of which he was a member, had completed the most dangerous part of its voyage – from Glasgow to Gibraltar – without incident. They had run the 2,000-mile Atlantic gauntlet without spotting either U-boats or Luftwaffe aircraft and the 750 miles they had steamed through in the Mediterranean had been equally uneventful.

Off to starboard, Brims could plainly see the Atlas Mountains of Algeria. They were less than 15 miles off the coast and he was convinced that in these waters he was safe from U-boat attack. Less than three furlongs from his port side, a British minesweeper – HMS *Byms* – was on station, and a glance in her direction boosted his confidence even further. A clear, sunny Saturday afternoon. What more could he want? Except, perhaps, to be at home in Liverpool watching his beloved Everton soccer team in action.

It was the last thought the Captain would have.

At 3.42 p.m., a torpedo struck the *Contractor* amidships, killing Captain Brims and three of his crew instantly.

Now it was the turn of another Captain to experience a surge in his spirits.

Kapitanleutnant Waldemar Mehl had a double reason to be happy. Today was his 29th birthday, and this 'kill' had brought his personal tally of Allied shipping sunk to over 25,000 tons.

He had joined the Kriegsmarine at 19 and, after six years' service aboard a series of battleships, had switched to naval intelligence at the outbreak of war. Two years later, seeking a suitable outlet for his devil-may-care personality and the need to experience real front-line action, he joined the U-boat service in 1941. Now a veteran of countless Atlantic patrols, he was commander of U-371 based in Toulon, home of the 29th U-boat Flotilla.

This 13th, and thus far lucky, patrol for U-371 had started on 22 July. Now, as he ordered his boat deeper and away from the British naval escorts already launching depth charges in the area he had just vacated, Mehl was calculating he had fuel for at least four more days at sea. He decided he would stick to the straits between Sardinia and Algeria – a regular passageway for Allied shipping. Content, as ever, in his work, the young Captain ordered a course change to northeast by east.

Unbeknown to Waldemar Mehl and to Douglas Lidderdale, who was now supervising the final lashing down of his Tiger on the deck of the *Empire Candida*, fate and the Captain's random choice of course had determined their rendezvous in 48 hours' time.

The *Empire Candida* left its berth in Bizerte on schedule at midday, rounded the northern headland of Cap Blanc, and was now steaming westwards at a steady seven knots with dusk approaching when Doug first realised that they might be facing a new danger.

He was on the bridge, talking with the First Officer, when the Captain drew their attention to a commotion going on to the northwest, some mile and a half off their starboard bow.

Doug lifted his binoculars and was able to see planes – which appeared to be Beaufighters from this distance – diving on a barely visible object on the surface.

The Captain had no difficulty recognising the RAF's target.

'It's a U-boat,' he announced. 'If the pilots hadn't spotted him, he might have been taking pot shots at us by now. Still could, of course, if they don't put him out of action.'

'What armaments have you got?' asked Doug.

'Armaments – virtually none. An anti-aircraft machine gun and anti-torpedo nets. We'll get those out but the gun's not much good at this distance, I'm afraid. Though, with our fellows buzzing about so low, we couldn't take the risk of firing anyway.'

'It looks as though the U-boat is firing back,' said Doug. 'They seem to have several ack-ack guns on the go. They have a big gun too but aren't using it.'

'That's the 88mm job,' said the Captain, binoculars still glued to his eyes. 'Same size as on your Tiger, so you know what kind of damage that can do, even from that distance.'

'Yes, I do,' said Doug. 'I know the effect only too well. It takes one to know one. Or to take one on, for that matter.' Then his eyes lit up with sudden excitement. 'I've got an idea.'

He grabbed the First Officer by the arm and pulled him towards the exit. 'I'm going to need some of your chaps to get the covers off a bit sharpish. But I think our Tiger is going to go hunting.'

By the time they reached the deck, he had explained his plan to the officer, who laughed out loud.

'I think you're quite mad,' he said. 'But what have we got to lose? A hell of a lot I imagine, if the RAF don't succeed.'

'Well, if my old chum from Maison Blanche, Joseph Berry, is up there, then I think we'll be all right. But there's no harm in having a bit of insurance.'

He spotted Baker enjoying the last of the sun on deck, and told him to find the two other crew-members. They were all present by the time the tarpaulin covers were stripped off the Tiger. Fortunately, everything had been left ready for driving the tank ashore at Bone so it was the work of just a few moments for Doug to check the gauges and fire up the Tiger's engine.

Leaving it running, he hauled himself up through the command cupola and fixed his binoculars on the distant U-boat. It appeared not to have suffered any damage, and had, he noticed, started to nose round to face in their direction.

'Christ, I think the crazy bastard is going to have a go at us even though he's under attack. Let's hope the hydraulic system works.'

Doug knew that, using the tank's hydraulic system, in less than a minute he could turn the turret the 90 degrees to the right needed to align the gun in the right direction. By hand, he would have to turn the traverse wheel 180 times to achieve the same effect.

After ordering his crew to put in earplugs, and muttering a silent prayer, he pressed forward on the pedal under the gunner's seat, then breathed a sigh of relief as the turret smoothly turned to order. On the other side of the cabin, Lance Corporal W.F. Pumfrey was in the loader's seat, inserting a high-explosive round into the breech.

The Leitz Turmzielfernrohr binocular gun sights gave Doug a two-and-a-half-times magnification of his target area, and a table of inverted 'v' signs on the right scope helped him judge the distance to the target, which he calculated at just under 2,500 metres.

When the gun was finally raised to the angle that satisfied Doug, he shouted a warning and fired. Even with the earplugs, the noise was tremendous and the Tiger lifted several inches, despite the heavy ropes lashing it to the deck.

Doug was already out of his seat and thrusting his head out of the command cupola, binoculars to his eyes, when the shell completed its four-second journey to the target. It missed by a good 50 yards but sent a huge waterspout hundreds of feet in the air.

The First Officer was jumping up and down on the deck with excitement. 'Bloody marvellous,' he yelled. 'Absolutely bloody marvellous.'

'Well, let's see if we can get any closer,' shouted Doug, a huge grin on his own face matching the sailor's.

He was making his way back to the gunner's chair when the First Officer called out again. 'I don't think you're going to get another chance. He's crash diving. I don't think he likes being fired back at. With the Beaufighters circling around upstairs and knowing what we've got for him waiting down here, I can't see him hanging around. There must be easier targets for him than this one.'

This was precisely the verdict of Captain Mehl. He would stay safely submerged and head directly for the shore-side comforts of his home base in Toulon. And what a strange tale he would have to regale his fellow U-boat commanders with in the mess.

Being attacked by a German Tiger tank from the deck of a helpless British merchant ship. He doubted anyone would believe him.

CHAPTER TWENTY-FOUR

10 AUGUST 1943

D oug had brought Tiger 131 as far as Bone. It was three-and-a-half months since its capture and he still had no idea how much longer it would take him to fulfil his promise to Winston Churchill and get the German monster back to 10 Downing Street.

He knew that his onward journey depended to a large extent on the main strike against Italy, which could not take place until the Sicilian invasion was successfully concluded.

The overall commander of this was the American General Dwight D. Eisenhower. But Churchill had again entrusted General Harold Alexander with the ground control of the war in Sicily, which had begun on 9 July. Monty had been made second in command despite his successes in North Africa.

On hearing of his reappointment as joint commander of the 8th Army in Operation Torch, Montgomery had remarked, 'After having an easy war, things have now got much more difficult.'

General Alexander had no sympathy for what he took to be Monty's pessimism. 'Back in Harrow, old boy, we'd tell you to lighten up.'

Monty looked at his counterpart. 'I'm not talking about me, I'm talking about Rommel!'

However, Churchill had always felt a little uneasy about Field Marshal Bernard Montgomery, even though he was one of his top commanders. He said to a Cabinet colleague, 'Monty exercises extreme caution which accounts in large part for his success, so long as he has a supremacy of manpower together with the backing of air power.'

Doug rationalised that he and his Tiger were held up in Bone because of preparations for an imminent strike against Italy. But he couldn't know that it was also creating a much more dangerous situation for him on the voyage home.

Smoking a cigarette and looking across the near-empty harbour of Bone in Algeria, Doug reflected on the happenings of the world stage during the last nine months and considered where he stood in the grand scheme of it all.

Since the beginning of 1943, despite the success of Alamein, the Allies were slow to gain the ascendancy. There were setbacks, but the trend was definitely in Churchill's favour. All of which did nothing for the dangerously brittle temper of Adolf Hitler who was the Commander-in-Chief of all German armed forces, including the Kriegsmarine. His authority was exercised through the Oberkommando der Marine (OKM), Nazi Germany's Naval High Command.

Early in 1943, Hitler appointed Karl Doenitz as Commander-in-Chief of the Navy. The naval headquarters was a huge communications bunker built 23 miles north of Berlin. It became known as the 'Koralle'. In the period when the Tiger 131 was being spirited out of Africa by Major Lidderdale, Adolf Hitler paid a visit to the Koralle.

Doenitz was loved and respected by his men, who called him

'The Lion'. During the First World War, Doenitz had served with distinction in what was then the Imperial Navy. In the Mediterranean, he commanded a U-boat – U-68 – that was sunk by the British who captured him and made him a prisoner of war. During his captivity, he came up with the idea of submarines hunting in packs, a practice the Germans called *Rudeltaktiks*. Later, when he had the opportunity to put his theories into practice, these feared groups of subs became known by the Allies as Wolfpacks, each pack numbering between 10 and 30 U-boats hunting as a coordinated entity.

Now, as he stood before Hitler, nobody could have conceived that Doenitz would hold the rank of President of Germany within three years, albeit for a mere 24 days. Doenitz stood rigidly to attention between two aides as Hitler strutted in front of him.

'The German Navy is utterly useless,' the Fuhrer spat. 'Your ships should be taken to the scrap yards. And the guns should be removed and used for coastal defences.'

Hitler paused to see what effect he was having on Doenitz. But the Admiral's face, though deathly pale, betrayed not a jot of emotion.

Hitler pressed on. 'The British have invented midget submarines that can attack big ships at will and then disappear into the night. Why can't we do that? Are you deaf, Doenitz? What is your answer? What is your strategy?'

Doenitz did not flinch. He spat out the facts with icy calmness. 'We are developing one-man submarines, Mein Fuhrer. But allow me to remind you that in the Mediterranean alone our U-boats have sunk twenty major Allied warships, including twelve destroyers, four cruisers, two aircraft carriers and one battleship, and up to ninety merchant ships – 400,000 tons of shipping.'

Hitler wrapped himself in his arms and swayed as he responded, spittle forming on his lips. As usual, with his phenomenal memory, he had the facts he wanted for his argument at his fingertips. 'We have lost 43 U-boats in the last month, 34 of them in the Atlantic. That is, 25 per cent of our entire operational strength. Few of the submarines in the Mediterranean have made it back to base.'

Doenitz felt compelled to put up a defence. He spoke more forcefully now. 'Since your favourite general, Marshal Rommel, lost El Alamein, the one branch of the Armed Forces to strike the greatest terror in the enemy is the Navy. Our success threatens to cut off their supplies.'

Hitler stepped up to the Admiral and shouted at him barely six inches from his face. 'Forty-three U-boats lost in one month. Explain that, Admiral.'

Admiral Doenitz was left reeling. 'I have ordered all war patrols to withdraw.'

'I don't permit you to run away.'

'This is not a retreat, Mein Fuhrer.'

Hitler exploded. The left side of his body trembled as he sprang forward. 'How dare you speak to me like that? If it's not a retreat, what do you call it?'

'It's an organised withdrawal on a temporary basis so that the submarines can be refitted with new and more deadly armaments.'

After each outburst of spittle-filled rage, Hitler strode up and down the carpet edge before suddenly stopping to hurl yet another accusation at the Grand Admiral.

'Your cowardly Feindfahrts will be out of action for months.'

'But, when they go back into action, Mein Fuhrer, they will have more powerful guns, better radar, and they won't have to surface to transmit radio signals. They will have the ability to communicate by raising their periscopes.'

It took courage to stand up to the Fuhrer, and Hitler secretly admired any man who did so. He stood staring at the Admiral, his left foot involuntarily tapping the floor.

Finally, he spoke. 'Very well. I'll give your Navy one more chance to prove itself. But you may as well know that my Tiger tanks are coming off the production line at an unheard-of rate. They will win the war, you will see.'

As in so many of Hitler's forecasts, which often relied heavily on astrological predictions, he was wrong.

For four months, most of the German U-boat fleet disappeared from the seas. The British Admiralty could hardly believe its luck. But, while holding its collective breath, the Admiralty never lowered its guard.

What Doenitz could not have suspected was that the British had broken the supposedly foolproof German 'Enigma' code communication system. The Enigma machine was an electromechanical device that relied on a series of rotating 'wheels' or 'rotors' to scramble plaintext messages into incoherent ciphertext. The machine could be set in billions of combinations, and each one would generate a unique ciphertext message. If you didn't know the Enigma setting, the message was indecipherable.

From Hut 8 in Bletchley Park, a picturesque village in Buckinghamshire, a group of mathematicians were now monitoring Doenitz's orders round the clock, as well as keeping an eye on the daily movement of his entire fleet.

Knowing that if the Axis powers gleaned the slightest suspicion that their ciphers had been broken the German cryptologists would invent a brand-new system, all the personnel working at Bletchley Park had to sign the Official Secrets Act. Part of that form included these injunctions:

Do not talk at meals...
Do not talk in the transport...
Do not talk travelling...
Do not talk in the billet...
Do not talk by your own fireside...
Be careful even in your Hut.

Cryptanalysts were selected for various intellectual achievements. There were linguists, chess champions, crossword experts and great mathematicians. Collectively, they became known as the GC&CS or the Golf, Cheese and Chess Society. At one time, the ability to solve a *Daily Telegraph* crossword in less than 12 minutes was used as an entrance test.

The accomplishments of Bletchley Park were top secret. Only a select few knew what was going on. Even most senior officers scratched their heads at the near-miraculous predictions that they received from intelligence on an almost daily basis. They referred to this information as coming from 'the Ultra decrypt department'. Winston Churchill referred to the Bletchley staff as 'The geese that laid the golden eggs and never cackled'.

The scrambled red phone in Churchill's office rang one evening. It was Roosevelt. Douglas Lidderdale had been correct when he surmised the red phone's purpose. In fact, there was another soundproof room opposite Churchill's office which was frequently mistaken for a toilet. A sign on the door read either 'Free' or 'Engaged'. It was from this booth that Churchill made outgoing calls to the White House.

'I want to say a big thank you from our intelligence people here,' said the ailing President.

Churchill was bemused. 'A big thank you for what?'

'Sharing the decrypting achievements of Bletchley Park.'

'It is common courtesy, Mr President, to share useful secrets with one's Allies.'

There was a thoughtful pause from the Oval Office end. 'May I be so bold as to ask why you're not giving Mr Stalin the benefits of the cipher-breakers?'

'Mr Stalin is a communist, Mr President.'

'Undoubtedly, but he is also our ally.'

'For the moment, sir. Only for the moment.'

Churchill was not to be shaken in his resolve to withhold all intelligence matters from the Russians. He was convinced that communists could not be trusted.

'But for the time being, Mr President, any man or state who fights against Nazidom will have our aid.'

CHAPTER TWENTY-FIVE
11 AUGUST 1943

Following the relentless bombing raids that wreaked apocalyptic destruction on Hamburg between 24 July and 2 August 1943, the cry for revenge by the German people could not be ignored. They wanted blood, and Doenitz, for one, was determined to give it to them. Once the U-boats had completed their refit with new conning towers bristling with two-barrelled 2cm 38 MII Guns, a 'Flakvierling' – a quadruple-barrelled gun – and improved radar, Admiral Doenitz ordered his submarines back to sea.

The date happened to be 12 August, a Thursday, the same day that Doug stood at a makeshift bar on a wharf in Bone on the North Coast of Africa. He was haggling over the price of a large scotch of dubious provenance and wondering where the hell the Liberty Ship he had been promised had got to. Flies were everywhere and the heat was sweltering. Eventually, he gave up and joined Lance Corporal Pumfrey outside the officers' refreshment tent.

'You can come in here if you like, Corporal.'

'No, thank you, sir. I know my place: the NAAFI.'

'Yes, but the nearest NAAFI is 20 miles away. There's only this tent.'

'That is why I am standing out here, sir, enjoying the benefits of a bit of sunshine.'

'Did you get the camouflage netting?'

'Sir!' The Corporal sounded faintly admonishing. 'I used to work with Sergeant Shaw. He taught me everything he knew. There are precious few things what I can't get my mitts on.'

'I didn't mean to offend your sensibilities, Corporal. Let us go and make sure our Tiger is hidden from prying eyes.'

The two men trudged across half a mile of desert to where the Tiger 131 stood between two sand dunes – not the most inconspicuous place to hide a 60-ton tank.

At close quarters, the war appeared to have been kinder to the small city of Bone than Doug remembered from his brief previous visit. What damage had been incurred on her wide tree-lined boulevards had been repaired in an astonishingly short time. There was a small cinema showing *Mutiny on the Bounty* and *Casablanca* on a regular basis and half a dozen small cafes that served croissants and sweet Muscatel at a shilling a litre.

The Regimental HQ of the tank unit was situated 20 miles south of Bone at Ain Mokra, where stones lined the neat gravelled paths and white flag poles were erected. The tank park of the 25th Infantry workshop, REME, was surrounded by barbed wire. Cans with pebbles inside them were attached at regular intervals to alert the sentries should the enemy try to purloin a 40-ton Churchill tank.

Leutnant Freidrick Kempka and his wingman, Oberfeldwebel Hermann Gunsche, had been scrambled with the other five planes of his squadron to seek out a lone Beaufighter reported flying off the Algerian coast. The two-engined Bristol

Beaufighter was a heavy-duty plane. If it was really flying on its own, it might have been a straggler separated from an Allied squadron – easy pickings.

After an hour of fruitless searching, the Focke-Wulf Fw 190s had failed to make any contact.

'Where are we?' said Kempka.

'Sardinia is a hundred miles northeast of us, Leutnant,' replied Gunsche.

Kempka checked his fuel and gave a crisp command to the squadron over the R/T. 'I saw a lot of activity in Bone harbour. Let's shake them up a little.'

The six Fw 190s gently turned starboard and approached the coast in well-practised formation. All Luftwaffe pilots liked the single-engined Fw 190. It was simple, fast and safe. Its only drawback was an inability to climb much higher than 10,000 feet. Kempka climbed to its ceiling and then dived down towards the small town of Bone.

Doug and Corporal Pumfrey had just finished pegging down the camouflage netting when Leutnant Kempka's planes swooped out of the sky and raked the silvery sand.

Doug could see the winking flames as their cannon fired. He didn't waste time communicating with his Corporal; he merely threw himself on to the sand, crawled under the netting and buried himself beneath the Tiger 131. Its virtually impregnable five-inch-thick armour would come in useful.

From the outskirts of the town, the heavy cacophony of ack-ack guns – anti-aircraft firepower – started up. Soon, though, the noisy engines of the Fw 190s drowned out everything.

'Break off. Break off. Follow me,' said Leutnant Kempka into his R/T.

On the horizon, he had caught sight of an airfield with an aeroplane on the runway; a plane with engines so huge they looked too big for it were fitted to each wing. Could it be the Bristol Beaufighter? Yes, it was. One after the other, the Fw 190s dived and strafed the plane. A wing fell off and an engine caught fire; the tyres burst, and at last the British fighter crumpled beneath the onslaught.

Satisfied, Leutnant Kempka pulled back on the stick and his plane zoomed high into the reddening sky, silhouetted against the setting sun. The squadron headed over the Mediterranean Sea on their way to their Italian base.

Doug crawled from under the tank and blinked. His light-green tropical kit was covered in mud. Pumfrey crawled out and stood beside him. He was dressed in khaki drill, and remained immaculate.

Pumfrey looked at his superior officer. 'Found a mud puddle to cool off in, did we, sir?'

The sarcasm was not lost on Doug, but he took it in good grace. 'Do you think they saw the Tiger?'

'Hardly. It's camouflaged well enough. From the air it must look like a sand dune.'

'I'm not so sure. Until we can get it on board a boat, there's got to be a better place to hide it than on the seashore.'

'I was under the impression we'd got over the worst part of our adventure, sir. I understood that, once we got to Bone, all we had to do was to load the damn thing on to a Liberty Ship called Ocean something-or-other and sail it home.'

'You may have noted, Corporal, that SS *Ocean Strength* is not yet in the harbour.'

'Strangely enough, I had noticed that glaring omission. But it is coming here, is it? I am right about that?'

'What exactly are you getting at, Pumfrey?'

'I was wondering whether it might have been sunk.'

'I think I would have been told if it was missing.'

The Corporal stroked his emerging moustache. 'I wouldn't put money on it, sir.'

'Anyway, they are more likely to sink it once we've got their captured Tiger on it, if ever they were to find out. The tank crews were under strict orders to destroy any abandoned Tigers to prevent them falling into our hands.'

'Are you suggesting that they'll stop at nothing to stop us getting to England, sir?'

'If they discover we have brought it to Bone, yes.'

'You don't think they know already?'

'How could they?'

'There are spies everywhere. Bone was in the hands of the Germans only a few months ago. We are probably being watched right now.'

'I wish you wouldn't always be so damnably optimistic, Corporal. Your cheeriness is getting me down.'

'I wasn't being cheerful, sir.'

'I know you weren't. I was being sardonic.'

'I see, sir; the privilege of being an officer.'

'Yes. And caviar to the General. Where the hell have Baker and our driver got to?'

'Baker and Wilkes said they were going on a recce to find a better hiding place for the beast.'

Doug looked sceptical. 'Any excuse to skive off, I suppose.'

'If you give me your uniform before you turn in, sir, I'll have it scrubbed up good as new by morning,' said Pumfrey.

'I don't have a batman, Corporal.'

'Don't worry about that, sir. I think I can handle it.' He broke off and said in an undertone, 'Be on your guard, sir. Someone's approaching.'

Pumfrey sprang into a defensive posture. 'Nobody knows that we're here.'

'That's what I thought too, but...' Doug stiffened as he watched the rapidly approaching figure. There was something familiar about the gait. Then he burst out laughing and ran forward to greet his old friend.

The newcomer quoted a favourite line from the *ITMA* radio programme: 'Can I do you now, sir?'

'Reg!' he exclaimed. 'I thought they'd sent you into Europe.'

'Not yet, sir. I came here the hard way – by road. I'm camped 20 miles south of here.'

'How did you know where to find me?'

'Colonel Taylor's in charge. Said you were hiding out somewhere on the coast. Then I came across your driver in town and he spilled the beans, so here I am.'

Reg Whatley shook hands with Doug wholeheartedly.

'So what are you doing?' asked Doug. 'You must have recovered all the old tanks by now.'

'Exactly. So I trudged along to see if I could lend you a hand. Mind you, I'm on permanent standby. Could be called away any minute.'

'We must do the town, old boy. There's got to be some sort of night life here.'

'Count me in. But I'm going to have to report back by reveille.'

At the far end of Boulevard Victor Hugo, as it neared the docks, there was a tatty-looking adobe building called the Club Exotique, which did not live up to its promise. Sailors and off-duty soldiers congregated here, and it was to this area that the ladies of the night tended to flock.

For want of finding anywhere better, that night Doug and Reg ventured inside the club, but the two old pals sat in a quiet corner and yarned of their exploits together.

When it was time for Reg to leave, he said he'd try to come back tomorrow.

'Please do try. I don't know how long I have to sit here waiting for a blasted boat. It's very unsatisfactory.'

'And you've parked the Tiger in a vulnerable place.'

'I'm not driving it all the way to the tank depot, if that's what you're hinting at. It's still got an overheating problem.'

'But disguising it as a camel under netting near a wadi on the edge of the desert is a bit precarious. Why don't you ask the French Consul if there's somewhere you can put it out of sight?'

'That's an idea.'

'Tell you what. If I can get back here tomorrow, I'll do it myself. First thing tomorrow, I'll make a call to the Consulate and arrange to pop in.'

'Right. That'll give you something to do.'

'Nice to see you again.'

Reg drove off in his Jeep. Doug trudged back to his tent on the sand near the Tiger. The midges feasted on him that night.

CHAPTER TWENTY-SIX
12 AUGUST 1943

In Africa, sunsets don't dawdle. One moment, the great red orb fills the horizon, the next it has disappeared to leave a canopy of sparkling stars as bright and clear as fairy lights on a Christmas tree, seemingly close enough to touch. Within minutes, the air turns from egg-frying heat to ice-cube cool.

Reg Whatley didn't manage to get back to the coast until nightfall the following day. He had taken to changing out of his tropical kit and into a thick battledress in the evenings. This was how he was dressed as he stood at a refreshment stall near the French Consulate in the town of Bone. French Boy Scouts served him lemonade to sip while he was observing the usual evening promenade.

The Frenchmen were immediately recognisable because of their immaculate suits and hats; most wore white, but some were more daring in matching grey or blue. The French ladies were universally elegant and their dresses reflected the fashion houses of Paris of the late 1930s.

There were younger women too, some very young indeed and often of mixed blood, who wore loose blouses and no brassieres.

They grew in numbers as the night wore on and all the respectable ladies had disappeared to their villas.

The Arabs, for the most part wearing white kaftans and red fezzes, often tried to sell tangerines or dates. Dates were eight pennies for a pound, the same price as twenty cigarettes. Reg preferred to waste his money on cigarettes.

With a few minutes to spare before his appointment with the Vice-Consul at 8 p.m., he lit a Gold Flake and started crossing the Rue Gota Sebti. Before he had taken two paces, he became conscious of a stunning young woman walking directly towards him from the direction of the Consulate. She smiled at him. He hesitated. She was looking at him as if she knew him.

Concealing most of her long golden hair was a large-brimmed hat that would not have been out of place at Ascot. A white silk dress clung to her body like a second skin, leaving nothing of her perfect figure to the imagination. Reg also noticed that she had the most delightful little turned-up nose. He reckoned that she could scarcely be much older than him. The young Lieutenant wrenched his eyes away and appeared to be studying the stork's nest on the adjacent mosque.

'Have you got a light, please?' asked the beauty in impeccable English.

Reg could not ignore her. He turned as she inserted a cigarette into a long holder. He fumbled for his lighter and tried in vain to control his trembling fingers as he lit her cigarette.

'The Arabs consider that the stork brings luck,' she said calmly with a smile.

He digested this, but couldn't think of a reply and was about to turn away when she spoke again.

'You shouldn't drink that filthy stuff.'

'It's only lemonade.'

'But you don't know what they put in it. Or in the stuff they call wine round here.'

'Quite,' he said. He coughed nervously, not wanting to be impolite. 'Would you excuse me? I've got to go and see someone in the Consulate.'

'You are here to see the Vice-Consul?'

'Yes.'

'Captain Blanchard.'

'Yes,' he repeated cautiously.

'I thought it must be you. I'm so sorry but he sends his apologies. He has had to go away very suddenly. I think you've got me instead.'

Reg looked bewildered. 'I'm sorry. I don't understand...'

'I am the Vice-Consul's wife.'

'Oh! Mrs Blanchard.'

'Madame Blanchard.'

'Of course. Forgive me.'

She laughed. It was a lovely, tinkly laugh. Reg found it, and her, beguiling. He smiled.

'I do believe you mistook me to be a lady of the night.'

Reg felt his cheeks turning pink. 'Oh, gosh. I'm dreadfully sorry. I had no idea. I thought...'

She stopped him. 'I know what you thought. How long have you been in Bone?'

'I arrived at the harbour in the last couple of days, actually.'

'Well, you weren't to know. I must introduce you to some of the ladies so that you don't make the same mistake again.'

Reg nodded. He was speechless with embarrassment.

She went on. 'I'm sure Captain Blanchard would send his apologies if he knew you were waiting for him. He cannot see you tonight. Unfortunately, he left the country this morning.'

'They didn't tell me that.'

'You probably spoke to one of the Arab interpreters.'

'Yes.'

'Yes. They are very cagey.'

Reg smiled at her choice of words. 'I've got to tell you, Madame Blanchard, you do speak English extraordinarily well. Much better than I speak French.'

Connie Blanchard gave another demonstration of her laugh.

Reg beamed. She aroused something in him.

'The reason I speak English so well is because I went to Roedean.'

'But your name...'

'Captain Blanchard is a Frenchman and I married him. English and French do sometimes intermingle.'

The Lieutenant looked sheepish. 'Oh, I say – I am a fool. Maybe it's because I haven't had a conversation with a woman for so long.'

Connie looked concerned. 'Oh, my poor boy, haven't you? We must do something about that.'

Reg grinned. He looked down at the sand and gave it a gentle kick. He was like a child confronted with his first knickerbocker glory.

'Where do you come from?' he asked.

'Southampton,' she responded immediately.

Reg's eyes sparkled even more. 'Good Heavens!'

Connie looked at him darkly. 'Don't tell me you come from there too?'

'Not exactly. But not a million miles away. The Isle of Wight, actually.'

'Oh.'

'I bet you've been there lots of times.'

'Oh, yes. Absolutely.'

The conversation seemed to have hit a buffer. After a

moment, all Reg could think to say was, 'Roedean College has been taken over by the Navy.'

'I didn't know that.'

'Yes. Now it's being used to give sailors torpedo training.'

'I wish they had sailors there in my day.'

They looked at each other, two lonely souls; mutually attracted no doubt, but without the words to express themselves.

'May I ask why you wanted to see Captain Blanchard so urgently?'

'It's rather a delicate matter.'

'Of course, if you'd rather not discuss it...'

'Not at all. But can we walk? I wouldn't like anyone to overhear us.'

'My goodness, this does sound exciting.' She put her hand on his elbow and guided him along the busy street away from the French Consulate in Rue Gota Sebti.   'Is all this stuff awfully cloak and dagger?' she asked, as they strolled.

Reg looked round nervously. He spoke in a low voice. 'Please don't repeat anything, will you?'

'You haven't told me anything to repeat yet,' she said gently.

'We have a tank.' He paused, not sure how much he could reveal to her, or even whether he should tell her anything. He only had her word that she was, in fact, Madame Blanchard.

'A tank? Yes, I've heard of those...' Her voice had a mischievous tone. 'As a matter of fact, there's one over there if you look.'

By the mosque, an abandoned wrecked Churchill tank lay on its side like a dead animal.

'Our tank is rather special. And we want to conceal it somewhere until we can get it on to a boat.'

'You want to hide your tank?'

'That's it in a nutshell ... Not that you could get a tank into a nutshell, of course.'

The tinkle he received in response was almost worth the crassness of his remark.

They walked a little further until they found themselves in the peaceful central park of El-Houria. They were alone among the lush tropical vegetation. Reggie's heart was racing. He wondered if hers was too. They looked at each other anew, lit by the bright moonlit sky.

'Are you married?' she asked softly.

He shook his head.

'You should think about it,' she said, then added thoughtfully, 'You could try the aerodrome.'

The non-sequitur threw him for a moment and he quickly dismissed the idea. 'It's not the safest place in the world. I watched it getting bombed this afternoon.'

'Exactly. What would be the point of bombing it again in a hurry?'

Reg remained dubious.

'It was only a dummy aerodrome, anyway,' she said.

'How do you mean?'

'It was bombed to smithereens over a year ago. Made completely inoperable. But we stuck a couple of planes back together and made it look as if we still use it.'

'I remember seeing a Beaufighter parked there,' said Reg. 'I heard that it was completely wrecked in this afternoon's raid.'

'My dear boy, it was just a shell.'

'You mean it was deliberately put there as a decoy?'

'Correct. It worked too. After all, the Luftwaffe didn't hit the town.'

Reg started to get interested. 'What's in the hangars?'

'A lot of holes in the roof. Nothing else.'

'I see. So what are you suggesting?'

'Hide your tank in one of the hangars. The chances are it'll be safer there than anywhere else.'

Reg laughed. He bought the idea.

'You have a nice laugh,' she said.

He looked at her wistfully. Something intangible, like a spark of electricity, passed between them. 'I must tell my senior officer about this.'

'And I must go home for supper. By the way, where is your tank at the moment?'

Reg hesitated and lied rather convincingly, 'I'm not absolutely sure what they've done with it.'

'Oh! Well, don't lose it.'

'Will I see you again?'

She took his hand and said gently, 'I do hope so, Lieutenant.'

'Reg,' he corrected her. 'Reginald T. Whatley.'

'And you must call me Connie.'

He repeated her name softly, 'Connie.'

She honoured him with a tiny tinkle of laughter as she sashayed back in the direction of the Consulate.

Lieutenant Reginald Whatley watched appreciatively and lit another cigarette. In his hand he discovered she had left him her calling card with a telephone number on it.

When he had finished his cigarette, he walked into a Café-Tabac, examined the card and made use of the phone.

CHAPTER TWENTY-SEVEN
13 AUGUST 1943

The following day, a Friday, was hot. By 1200 hours the temperature had climbed to 95 degrees Fahrenheit. Doug Lidderdale stood on the edge of the docks in Bone and lit a cigarette. Lance Corporal Pumfrey waited a respectful yard behind him, his hands clenched together at his back.

'I've left young Baker and Wilkes to guard the Tiger, sir.'

'Good.'

In an attempt to make conversation, the Corporal said, 'I had to get Wilkes to clean the barrel of his Sten gun. Bloody sand. Excuse my French.'

'Well done.' It was evident that Doug's mind was on other things.

Pumfrey was not deterred. 'God alone knows what they teach them in basic training these days. He asked me why it was called a Sten gun.' The Corporal laughed in mock derision.

Doug turned and stared at Pumfrey. 'Well? Go on. Why is it called a Sten gun?'

Pumfrey stopped laughing. 'I thought everybody knew that, sir.'

'Well, I don't.'

Pumfrey seemed shocked. 'It derives from the initials of its inventors' names – Shepherd and Turpin – and then the first letters of Enfield, where the gun was originally made.'

There was a long pause as Doug digested this. 'What perfectly useless information.'

'Yes, sir.'

'It's just as well the gun wasn't invented by Seymour and Hicks and made in Italy.'

'Yes, sir.'

Doug blew out a long plume of smoke that hung heavily on the hot steamy air. He turned away and stared over the docks. 'What's the bloody date?'

'The thirteenth of August, sir.'

'I might have guessed. Don't tell me it's a Friday?'

'Yes, sir; all day, sir.' Pumfrey's humour was of a limited and juvenile kind.

'Friday the thirteenth. That explains everything.'

'Explains what, sir?'

'It's our unlucky day.'

'It was my understanding that the Admiralty was laying on a Liberty Ship for us, sir.'

'That was what I gathered too.'

'Well, if you ask me, sir...'

'I don't.'

Pumfrey ploughed on regardless. 'In my opinion, it's a diabolical liberty, sir.'

Doug's grey-blue eyes looked pained. 'Oh, Corporal,' he said, 'do you say these things deliberately to wound my sensitive soul?'

'Couldn't resist, sir. But, on a serious note, what exactly is a Liberty Ship, should we ever have the opportunity of tripping over one?'

'It's a ship assembled along mass-production lines.'

'Made in America?'

'And Canada too, I believe. They're churned out like Ford motorcars. They are light enough to carry exceptionally heavy loads and still have a fair turn of speed. The only fly in the ointment is the welded construction of their prefabricated parts.'

Pumfrey nodded wisely. 'Prone to fracture easily?'

'Exactly. Dozens of Liberty Ships have sunk even when they haven't received direct hits.'

Pumfrey received this information with a faraway look on his face. 'And they want to send us back to England on one of these things carrying over 60 tons of tank?'

'Comforting, isn't it?'

'I suppose the Admiralty knows what it's doing, sir.'

Doug gave the Lance Corporal a querulous look. 'Do you really think so?'

Both men, so close to each other in age, but so different in their social ranking, stared out across the docks. They held their breath against the stink of stale fish and the acrid diesel as they peered out over the harbour, which was sparsely filled with small dhows and landing craft.

Lieutenant Whatley marched towards them at a fair pace. He came to attention and saluted.

Doug saluted back. 'You're sweating.'

'I've been looking everywhere for you, sir.'

'Well, you've found me. And I'm not going anywhere. None of us is.'

'That's just it, sir. You don't know how long you'll be stuck here. So I've been wondering where to hide the tank.'

'We've already hidden it.'

'I know, sir. But anybody could stumble across it. And you

can't trust anyone around here. There are a lot of German sympathisers and Vichyites.'

'Are you getting round to telling me that you've found a better hiding place?' asked Doug.

'Yes. The aerodrome.'

Doug and Pumfrey exchanged glances. They didn't have to say the word 'doolally' but it was obvious they were thinking it.

Reg continued. 'I was put on to the idea by the French Consul's wife – a very sweet lady. Very ... oh, what's the word I'm searching for?'

Doug tried to help. 'Convex?'

'Oh, yes,' said Reg dreamily. 'Very convex. And jolly helpful. Apparently, the aerodrome hasn't been operational for at least a year, but they make it look as if it is so as to draw air attacks away from the town. Clever wheeze, what?'

'Yes, I was here yesterday when the Focke-Wulfs mowed it down,' said Doug.

'The town was left untouched though. Wizard, eh?'

'Are you suggesting we put our tank in the one place where the Luftwaffe strafes on a regular basis?' said Pumfrey thoughtfully.

'In a hangar where they won't see it.'

'You reckon they won't attack a hangar?'

'Even if they do, the Tiger can withstand their piddly little bullets. And the beauty is, none of the locals will go near the place for the very reason that it draws enemy planes to it like a magnet.'

Doug laughed. Pumfrey glanced at him with concern.

'By God, I think it's a wonderful idea,' said Doug. 'But don't we have to get permission? Who owns the aerodrome, Reg?'

'The French authorities, sir. This is one of their colonies. *Algerie francaise* and all that.'

'So, who do we have to see?'

'Captain Blanchard. He's out of town at the moment. His wife can get us permission from one of his aides.'

'Written permission?'

'I can get it for you if you want to go through with it, sir.'

Doug turned to Pumfrey. 'What do you think, Corporal?'

'The desert sand is doing our Tiger no good at all. It's getting into every nook and cranny. It would certainly be easier to maintain in a hangar.'

'I agree. Very well. We'll move it under its own steam by cover of darkness. We'd better go and tell Baker and Wilkes. One of them is on guard duty at the moment.' Doug grinned at Reg. 'Well done, Reggie. Quite a brainwave.'

The junior officer basked in the praise. 'I'll go back and get the Consulate's stamped approval.'

'Can I come along with you, sir?' asked Pumfrey. 'Haven't seen a decent-looking bint for months.'

'None of that language, please, Corporal. This is the wife of the French Consul. You treat her with respect.'

'Yes, sir.'

'But do come along. I've an appointment with her. She's getting the authorisation for us now.'

'After that, I'll take the Jeep to Regimental HQ and see if there's any mail.'

'Thank you, Corporal,' said Reg.

Doug voiced his thoughts: 'An aircraft hangar, eh? Yes, the idea's growing on me.'

Reg and Pumfrey left together and Doug trudged towards the sand dunes.

Connie Blanchard sat sipping a glass of iced water in El-Houria Square. Sitting on a bench, her parasol held up against the sun, she looked as pretty as a film star. Lieutenant Reginald Whatley marched up to her accompanied by Lance Corporal Pumfrey. There were dark perspiration patches on their uniforms.

Connie put down her glass and rose to greet them. 'I've been waiting for you, Reggie.'

'Mrs Blanchard...'

'Connie, please.'

Reg hesitated, stunned momentarily by her beauty. 'Connie.'

Pumfrey was aware of the invisible spark between them, though his face barely betrayed anything other than mild amusement.

'This is Lance Corporal Pumfrey. He is one of our mechanics.'

Pumfrey gave a cursory salute.

Connie inclined her head as she appraised him. 'I was expecting to meet your commanding officer.'

'Major Lidderdale hopes to meet you soon. He has other things to do today.'

'Have you lots of things to do too?' Her voice was enticing.

'No, I have come to say we've made a decision to park our tank in the safe place you suggested.'

'Good. It does make sense.'

There was another lull in the conversation.

Pumfrey, not insensitive to a romantic atmosphere, decided to attempt a diplomatic getaway. 'Madame, if you can furnish me with the written authority, I will be on my way. As the Lieutenant has some time on his hands, I'm sure he would welcome the opportunity of seeing the sights. If you have nothing better to do yourself, of course.'

Connie opened her handbag, took out an official-looking document and handed it to Pumfrey. He gave it a cursory glance and tucked it into a breast pocket. 'Thank you, ma'am.'

'It's got the Consul's stamp on it. My husband's away.'

'I understand, ma'am. When do you expect him back?'

'Oh, not for some time. It depends when the conference is over.'

'Conference, ma'am?' Pumfrey expressed ignorance.

'There's a top-level conference in America. Codename Quadrant.'

'That's news to me,' said Reg.

Connie's eyes widened. 'Perhaps I shouldn't have said anything. Roosevelt has called a meeting.'

'Oh! Well, Mum's the word. Your secret's safe with me.'

Connie gave her laugh.

Reg grinned and looked at Pumfrey. 'If you're going back to pick up supplies at the depot, you'd better be off.'

The NCO saluted, spun on his heel and walked smartly away.

Connie flashed a smile at Reg. 'There are some very beautiful views round here. Would you like to come with me? I can show you some of the sights?'

Reg was hooked. With her, he would have gone anywhere.

Some two hours had passed by the time Pumfrey got back from his trip to the REME tank depot. He approached Tiger 131's hiding place on foot. 'I've left the Jeep by the wharf,' he said.

'Good,' replied Doug, looking up hopefully. 'Any post?'

'Nothing, sir. Just a note from a bloke in the British Council.'

Doug frowned. 'What's the British Council? Are you sure you don't mean the British Consul?'

'No, sir, Council. Here's his card.'

Doug stared at the small visiting card. Printed on it were the words: 'Dudley Wrangel Craker. British Council, Cairo', followed by a telephone number that had been crossed out.

Doug looked puzzled. 'Dudley Wrangel Craker! Are you sure this isn't some sort of joke?'

'Dressed very posh, he was,' said Pumfrey. 'White straw titfer. Bow tie. White whistle and flute. Tanned face. Looked like a negative.'

'You saw him?'

'Briefly. He was looking for you.'

'Did he say what he wanted?'

'Said he had to talk to you urgently. Wouldn't talk to an NCO.'

'I suppose he might have come from Cairo. There are a lot of operations going on there.'

'He said he had been in Egypt before the war, but now he has got an office in Rue Ali Biscuit or somewhere like that.'

'In where?'

'A country pile here in Bone.'

Doug looked blank.

Pumfrey went on. 'He scribbled the address on the back of the card.'

Doug read the visiting card. 'Rue Ali Biskri.'

'That's where the big hotel is,' said Pumfrey.

'How do you know?'

'I was thinking of billeting there myself. But it's a bit expensive, even for me.'

Doug gave a slightly helpless shake of his head. It became clearer by the day that a lot of Sam Shaw had rubbed off on his protégé Pumfrey.

'Come on. Let's see what Mr Craker wants.'

They arrived together at the Hotel Seybouse and went inside. It was the largest hotel in Bone and had all the colonial trappings: potted palms, ceiling fans and ornate statuary.

'This is the number on the card, sir. The person you want must be staying here. If you'll excuse me, I'll leave you to it.'

Doug took Pumfrey's arm. 'No,' he said earnestly. 'Please stay. I don't know who this man is or what he wants. Back me up, will you?'

'Certainly, sir.'

Immediately and impassively, Pumfrey followed his commanding officer to the little reception desk where an Arab in a white dress and a red fez salaamed at them.

'I'm looking for a Mr Wrangel Craker,' said Doug.

The gentleman behind the desk immediately indicated a room further along the ground floor, then smiled and turned back to his books.

Doug caught Pumfrey's eye. 'Come on.'

The room the clerk had indicated was a sort of lounge and refreshment area. In deference to Islam, no alcoholic beverages were on display. Doug's heart sank.

There was only one person sitting at the bar, if that is what it could be called. He wore an immaculate white suit and he had placed his white Panama hat on the counter.

Pumfrey tipped the wink to Doug who moved to stand next to the stranger.

'Excuse me, sir. Are you by any chance Mr Wrangel Craker?'

The man from the British Council gave a cursory glance at Doug and said, 'Glad you could make it. Call me Dudley, by the way.'

They shook hands.

'I'm Douglas Lidderdale and this is Lance Corporal Pumfrey.'

'Yes, we met. Sit down, Corporal. Sit down, Douglas. What can I get you to drink?'

'A lemonade would be very welcome,' said Doug cautiously.

'Lemonade!' Dudley spat out the word as if on the verge of

apoplexy. 'Have a Singapore Sling. Perfect for this weather. What are you having, Corporal?'

'I don't suppose they serve beer here?'

'You can get anything you want in a Muslim country so long as you are prepared to pay for it and drink it out of a cup. Here it is very much like the old Prohibition days in America.'

The men smiled and a barman appeared.

'Ah, there you are, Ali; two Singapore Slings and a bottle of Watney's, please. When the drinks arrive, we will sit over there, if you don't mind.' Dudley indicated the veranda. 'Even the walls have ears, and they're often bilingual.'

'You're from the British Council, I understand?' said Doug.

'I work for the British Council, yes.'

'I'm sorry. I'm not really familiar with...' Doug tapered off.

Dudley was only too willing to furnish the information. 'We are a worldwide organisation, Douglas. I hope you don't mind if I call you Douglas?'

Whether he minded or not was academic because Dudley ploughed on expansively. 'Our aim is to spread British Culture and Arts to the less civilised world. We arrange exhibitions and ballets; operas and stage plays. We do anything that shows off our prowess and attitude to the wilder, less privileged countries. I was based in Cairo before the war, then in neutral Spain, if you can believe it is neutral.' He rolled his eyes back in a rather extravagant manner.

Doug realised that the man from the British Council was possibly not wholly heterosexual.

Dudley flashed a charming smile at Pumfrey. 'Why don't you take your beer into the garden, dear boy? You'll find some really exotic Bougainvillea out there.'

Pumfrey took the hint and disappeared into the back garden, leaving the older men to their conflab.

Dudley went on flamboyantly. 'Nowadays, I see myself as more of a roving ambassador. But I still try to do good work and keep my ear to the ground, if you know what I mean. Let's move.'

Doug and Dudley took their drinks to the wide window by the veranda where they sat round a bamboo table.

'Don't worry about the gargle,' said Dudley. 'They allow me to run up a tab here.'

He looked round to make sure they were quite alone, then went on in a hushed voice. 'We won't stay together too long if you don't mind. Somebody might notice and start to wonder what we are jabbering on about.'

There was a pause while they sipped their drinks.

'Why exactly are we meeting?' asked Doug.

'A message has been passed down to me, through official circles – all hush-hush, you understand – that you can't have your boat for at least another month.'

Doug banged down his cup forgetting that it came with a saucer. 'But we were promised...'

'I know you were, old chap. I'm in on the whole thing. But the fact is –' he tapped his nose in a conspiratorial way like a villain in a silent movie '– there are many more plots afoot.'

Doug gave the other man a shrewd look. 'Do you mean the Quadrant meeting?'

Dudley Wrangel Craker stared in shocked amazement. 'How the devil did you know about that?'

Doug coloured. He had evidently stumbled on a touchy subject. 'Oh, Corporal Pumfrey heard some mention of it on the grapevine.'

'Good God! It's top secret. The conference only began yesterday. If a word of this gets out, it puts the top brass in terrible danger.'

'Not a word, sir. Not a word, I assure you,' said Doug.

Dudley took a couple of gulps to regain his composure.

Doug continued, 'So the conference is not the reason, I take it, for the delay in getting a boat?'

'Actually, there are several reasons. U-boats have been fairly inactive for the past three or four months while they've been fitted with better armaments. Now Admiral Doenitz has ordered his submarines to go out on patrol again. So the U-boats present us with a new, unknown danger.'

'You seem well informed of the German movements, sir.'

'Call me Dudley, for Christ's sake. And, yes, I am well informed because there's a group of boffins in England who we call Ultra that can tell us everything the Germans are going to do before they know themselves.'

Doug looked suitably impressed. 'Who has cracked the ciphers?'

Dudley shrugged. 'Did I say anything about ciphers? Perhaps Churchill is employing mind readers. Oh, Christ! Look, that's hush-hush too. The point is, until we can be sure where the U-boats will be positioned, we can't guarantee you a clear passage back to Blighty.'

Doug nodded sceptically. 'And that will take a whole month, will it, Dudley?'

'There are other things happening. Now, for God's sake, no one, I repeat, no one must get wind of what I'm telling you.'

Doug nodded and glanced round nervously.

Dudley's voice lowered to a near whisper. 'The Allies are about to invade the continent from the south. We don't want you and your tank getting in the way. The invasion is too important and so is the Tiger. So I'm afraid you're stuck here until the Italian campaign is under way. Can't have the Mediterranean cluttered with too many ships at once. After all, we have to give you an escort, and we're going to have to wait until some destroyers are freed up for you.'

Ali approached the table. The two men looked startled as the barman said, 'Can I serve the honourable gentlemen with another refreshing libation of an alcoholic nature?'

Dudley drained his cup, picked up his white Panama and said, 'No, thank you, Ali. Time for Tiffin.' He turned to Doug, who had also risen. 'Have a good weekend, Major. Hope you find something to occupy your mind while you're waiting. I'll be in touch.'

Dudley, weaving slightly, departed a shade theatrically, waving his hat in a gesture of farewell.

At the bottom of the garden, dutifully out of earshot, Pumfrey was smelling the flowers. Doug waved for him to return and they left together.

'I wonder if we can trust him, Pumfrey,' said Doug, as the men walked past the docks towards the open desert. 'After all, he didn't show us his credentials.'

'No, sir.' Pumfrey considered the matter and came up with a layman's conclusion. 'But if you think about it, if he really is a spy, credentials are the last thing he is likely to show to anyone.'

'True, Pumfrey. Ever thought of becoming a philosopher?'

'That's exactly the job I've applied for once the war's over.'

'I can see you right up there with Plato and Bertrand Russell.'

By the time they had returned to Tiger 131, the sun was a huge apricot on the horizon. They clambered beneath the camouflage netting and nearly fell over Reg, who was outstretched on his back by the tank tracks.

Pumfrey grimaced and turned to Doug, 'The officer is in love.'

'"Lieutenant" to you, Corporal,' said Reg.

'Lance Corporal, if you don't mind.'

Reg ignored him and turned to Doug. 'So when do you start mustering?'

Doug lifted an eyebrow. 'Mustering?'

'When are you gathering a convoy together? You're not setting sail solo, are you? Paddling your Tiger back to Blighty on a raft?'

'Reg, in the course of a war, priorities are bound to keep changing.'

The Lieutenant smiled. 'You know something, don't you, sir?'

'Even if I did, I wouldn't tell you.'

'There's a rumour that we're about to invade Italy.'

'Is there?'

'If it were more than a rumour, it might explain the lack of available ships.'

'Where did you hear this rumour, Reg?'

'From Madame Blanchard.'

'The lady seems remarkably well informed.'

Pumfrey intervened. 'She also told you that Roosevelt has called a meeting, didn't she?'

'Yes. She is in the know about a lot of things. After all, she is the French Consul's wife.'

'So long as she doesn't talk about every secret she picks up,' muttered Doug.

Reg stiffened. 'She confided in me as an ally, a compatriot.'

'Of course she did. But how many other compatriots does she have?'

Reg's face clouded. He got to his feet and as he spoke there was an edge to his voice. 'She's a damn fine woman, Major. She went to Roedean.'

'Before you blow your top, Reggie, may I remind you that she is also a married woman.'

'Her husband is at least 20 years older than she is.'

'What's that got to do with it?'

'Well, I mean – she's young,' stammered Reg.

'Yes. So I gather.' Doug gave him a reassuring smile. 'I'd like to meet her.'

Reg seemed mollified. 'I think that can be arranged, sir.'

Doug's expression was as enigmatic as that of the Sphinx. 'Good,' was all he said.

Reg replaced his cap and saluted. 'I'd better be getting back to the tank depot. See you tomorrow.'

'I hope so.'

CHAPTER TWENTY-EIGHT
14 AUGUST 1943

The mornings started off frosty and at 0630 hours, under the awning that sheltered Tiger 131, Doug warmed his hands round the mug of hot tea that Driver Wilkes had rustled up. With the rising sun, the warmth of the Sirocco wind swept over them from the Sahara desert in the south.

'I'm not convinced we should put the tiger in an airfield,' said Doug. 'I'm starting to wonder whether we should move to the comparative safety of the tank compound.'

'Back at the REME workshop, do you mean, sir? Blimey, it's 20 miles away. If our boat ever comes in, it'd take us all day to get the Tiger back here. And that's assuming the bloody thing doesn't break down.'

'We are still a part of a REME unit,' said Doug.

'Yes, sir,' said Wilkes. 'But we are also a team of four men within a unit in charge of the Tiger tank. At the depot, they have a habit of changing the Regimental CO every few weeks. We might be landed with a tricky bugger who doesn't understand our situation.'

Doug sighed. He knew Wilkes had a good point. He decided to wait for events to unfold for a while longer.

The moon was bright and it was not until late that evening that there was sufficient cloud cover for Doug and his crew to lift the camouflage net and drive the Tiger one-and-a-half miles to the defunct aerodrome. In one of the hangars furthest from the potholed runway, Pumfrey had prepared a straw-covered floor on which their Tiger could rest its caterpillar tracks.

Pumfrey was only a couple of years younger than Doug, but he had quickly fallen into the position of being Doug's unofficial batman. He had secured bunks for the four of them – nobody had the courage to ask from where – and put them into the two small offices that had once been occupied by engineers at the back of the hangar. These had been quickly converted into reasonably comfortable billets. Wilkes, having driven the tank without a hitch, turned in at 2230 hours.

Baker took the first four-hour watch, while Doug and Pumfrey left the airstrip and headed for town. By mutual consent, they had decided they needed a drink.

The Club Exotique had not improved since Doug's last visit. The ground floor was approximately 35 square feet. The mud walls had been whitewashed but not in recent memory. Half a dozen tables without cloths had been scattered around the place with a large assortment of ill-matching chairs. A bar, rather similar to the one in the Hotel Seybouse, was situated on one side of the room and the steward was a surly Frenchman with pretentions to being maitre d' at the Ritz. When one ordered a drink, he insisted on seeing the colour of one's money first. Arabs weren't barred, but they weren't welcome either. They seldom drank, but when they did they were liable to cause trouble.

In one corner of the room, farthest from the bar, an ancient upright piano – or 'Joanna', as Pumfrey called it – leaned against the wall. It had yellow ivory keys and some of its teeth were missing. The upper surfaces of the woodwork were so raddled with cigarette burns that the piano was beginning to look like a failed experiment in pointillism.

Doug bought himself a double scotch on the rocks and gravitated to the piano where he sat on the rickety stool and started to play 'Ain't Misbehavin'', a cigarette hanging lazily from the corner of his mouth. Pumfrey, nursing a bottle of Guinness, put an elbow on the piano and watched him appreciatively.

Major Douglas Lidderdale was a good player, a very good player. The dozen other customers in the bar stopped talking for a minute to listen to him. Some of the lowlife ladies listened even longer than that.

'I didn't know you could play, sir,' said Pumfrey, smiling.

'I always wanted to be in a jazz band. How I turned into a mechanical engineer I'll never know.'

At that moment, a dapper middle-aged gentleman in a white uniform approached the piano. He stood, listening appreciatively, his hand on the shoulder of his companion – a dark-haired native boy of about 13. Doug looked up at the newcomer. Ash fell from his cigarette on to the floor.

He brought the piece to a premature finish and stubbed out his cigarette on top of the piano. 'Are you the police? Was I playing too loudly?'

The man in the uniform gave a laugh that would have made Maurice Chevalier proud. He spoke with a heavy French accent. 'No, m'sieur, I am not a gendarme. I am the local Consular official.'

Doug extended his hand. 'Glad to meet you. I've been waiting to meet your superior. I have some private business to discuss. But this is neither the time nor the place.'

The Frenchman looked surprised. 'My superior? And who is that, if I may ask?'

'Your boss, Captain Blanchard, the Vice-Consul.'

'But I am Captain Blanchard,' said the Frenchman, nonplussed.

Doug sat back on the piano stool. 'Oh! I do beg your pardon. I thought you were in America.'

Captain Blanchard looked even more perplexed.

Pumfrey stepped in. 'I must have misunderstood what your wife was saying to me today.'

'My wife?' The French Vice-Consul looked aghast. 'But I am not married.'

Captain Blanchard's young companion giggled.

Doug and Pumfrey left Captain Blanchard in the Club Exotique and jumped into their Jeep. If there had been a timekeeper to record it, Pumfrey would have broken speed records on his drive back to the aerodrome. The cloud had rolled in thick and by the early hours the night was pitch black. The wide doors to the hangar were closed.

'Baker!' Doug called out. 'It's me and Corporal Pumfrey.'

There was no response.

'It's Major Lidderdale. I'm coming in.'

Both men put their backs into pushing the sliding doors and eventually managed to ease one of them open.

A man leaped out of the black vacuum. Doug could only see him dimly. Before he could shout a warning, the stranger had clouted Pumfrey's skull with something hard and he went down like a sack of potatoes. Doug rushed at the unknown assailant but missed and went crashing to the ground. He had tripped over Baker's inert body, which lay just inside the hangar's doors.

The intruder stood over him with a Luger in his steady grasp. He raised his foot and moved to kick Doug in the face. Doug shied away instinctively, but the boot connected with his ear, deafening him and leaving his brain reverberating like a jelly in his skull. Yet, somehow, the jar had helped him recover some night vision. At last, he could make out a vague outline of the man above him.

The man raised his foot to kick again but Doug spun round, took hold of the foot and twisted it viciously. He heard the ankle bone click. He couldn't have broken it, but some joint had come out of its socket. The man screamed and pulled the trigger of his Luger. Doug felt a bullet graze his cheek.

He leaped to his knees and grabbed the hand that held the gun. He yanked down until the man was also on his knees. Doug head-butted him rapidly – forehead against nose – three, four, five times.

The man cried out: 'Lieber Gott!'

Doug heard the clatter of metal as the Luger fell to the concrete. Blood poured from the German's nose into Doug's eyes. Temporarily blinded, he managed to get both hands round the man's neck. The man grabbed Doug's throat in return. Barely able to breathe, Doug used his last reserves of oxygen to force his body forward against the man, pushing him backwards. Before he passed out, Doug's only hope was to topple him on to his back.

The veins stood out on his temples and Doug's palms were slick with sweat as he put his last ounce of effort into forcing the man back. With superhuman effort, Doug managed to draw one of his legs back and knee his opponent in the groin.

The German screamed, overbalanced and toppled backwards

with Doug on top. Now, he had a momentary advantage over his weightier adversary.

What had they taught him on that weekend unarmed-combat course at Hilsea Barracks? *Thumbs into the arteries. Thumbs into the arteries.* Doug pressed the man's neck with all his might and gradually the grip on his own throat slackened. The man's eyes opened and rolled upwards in their sockets. His body slackened and collapsed.

Doug wiped the blood from his eyes while he fought for air. Straining for breath, he coughed and vomited. He tried to relax his fingers but found they were frozen tight like dead claws. Trying to open them caused excruciating pain.

The man's eyes began to focus again. Doug forced his fingers and thumbs to unfreeze and open centimetre by centimetre. Wasting no more time, he felt on the floor for the fallen Luger. Holding it by the barrel, he smashed the butt down on the man's left temple. The man's head spun sideways and he lay deathly still. Wilkes stumbled across to the doorway from the back of the hangar, carrying an oil lamp that illuminated the scene like an 18th-century painting.

'Have you killed him?' said Wilkes, still half-asleep.

'Not quite. Where the hell do we put prisoners of war?'

'He's not a soldier. He's wearing a boiler suit.'

'I'm pretty sure he's a soldier of some sort. This is a service revolver. Get some rope or wire or something. Quickly.'

Doug kicked the man on to his back, then knelt down and went through his pockets. There was some foreign currency amounting to about seven pounds and ten shillings, a folder of matches from the Club Exotique and an extra clip of cartridges. But there was no means of identification.

Wilkes returned with some thick fuse wire and they started to bind the man's hands behind his back. They were not gentle.

The wire cut into the man's wrists.

'Is he wearing an ID tag?' asked Wilkes.

Doug ripped open the top of the boiler suit and around the man's neck was a gold chain and a German army metal tag. There was a small skull embossed on it and, underneath, the man's service number and the legend 'SS OKH'.

'An SS Man. Himmler's lot. Bloody ruthless.'

Baker groaned as he began to regain consciousness.

Doug went to his aid and helped him sit up.

'What the hell happened, Major?'

'You got clobbered.'

'Oh yes, I remember – I was on guard duty. There was a knock. I thought it was you.'

'We were set up.'

'Set up? I don't understand. Set up by whom?'

'By a woman, I suspect. A woman calling herself the wife of the local Vice-Consul.'

Baker was holding his head and trying to focus on what Doug was telling him. 'You haven't met her?'

'No, but I have met Captain Blanchard. He's a lifelong bachelor.'

Baker held his head again and rose uncertainly to his feet.

'Can you stand all right? Good. Give me a hand with Pumfrey.'

Pumfrey had started twitching back to consciousness. The first person he saw when he came to was Doug. 'What hit me? A sledgehammer?'

'This Luger. We were all lucky it was so dark. If he could have seen properly, he'd have shot the lot of us.'

The following day, Sunday, Wilkes woke the other three members of the team with steaming hot mugs of coffee and plates of unidentifiable fried fish.

'What the hell is this?' said Doug after a mouthful.

'Local fish. Caught this morning.'

'Was it caught or did it commit suicide?'

Though Doug's team had thick heads and lumps and bruises the size of goose eggs, seven hours' sleep had revived them sufficiently to think out a strategy.

The German proved to be recalcitrant in any language so he had been driven in the early hours to the tank depot 20 miles to the south and locked in the cooler usually reserved for drunken soldiers. Lt. Colonel J.F. Taylor would deal with him on Monday morning, and Doug and his men would be on hand to give testimony.

Doug saw the Adjutant and requested an immediate meeting with Brigadier Noel Tetley, the 25th Brigade's latest commander, appointed the previous month.

The Adjutant said, 'He's conducting an interview at the moment, sir.'

Doug had learned with bitter experience the delaying tactics of senior officers. He was not prepared to sit on a bench for the rest of the day, reading ragged editions of *Lilliput*.

'This is most important, Captain. As you know, I've brought a prisoner in and there are implications.'

The Adjutant could see that Doug meant business and went to the door of the inner sanctum, on which he knocked timidly.

'I told you I can't be disturbed,' a voice barked.

'Urgent business,' said the Adjutant. 'Officer from the field, sir.'

There was a strange expostulation from within and then the door opened and Dudley Wrangel Craker walked out.

Doug stared at him.

Hidden within the room, Brigadier Tetley said loudly, 'Forgive me, Colonel. Must take news from the front.'

Doug continued to stare at Dudley who ignored him, tipped a sort of salute to the Adjutant, and sauntered out of the office.

Tetley shouted out, 'Well, are you coming in or aren't you?'

Doug entered and saluted.

'May I ask you, sir, do you know that gentleman?'

'Colonel Clarke? Of course I do. What do you want?'

'Well, this is rather awkward, sir. It's about that gentleman that I came to see you, really.'

'Oh, yes?' The 45-year-old Brigadier looked suspicious. 'What is your point, Major?'

'You say his name is Colonel Clarke, but he is not in uniform.'

'Even I wear civvies sometimes.'

'What then, may I ask, does he do, sir?'

'War correspondent for *The Times* now.'

'Not the British Council representative for Egypt?'

Tetley sniffed. 'He may be. Did he tell you that?'

'Yes, sir.'

'That's what he is, then.'

Doug stared hopelessly at the Brigade commander, who relented to a degree.

'Lidderdale, he is an intelligence officer. Do you understand? I suggest you ask no more questions.'

'I see, sir.'

'Good. I suggest you wipe Colonel Clarke from your memory. Now, anything else that you so urgently wanted to see me about?'

'I've brought in a German prisoner, sir.'

'Yes, I know. He will be questioned, of course.'

'Thank you, sir.'

'Well done, Lidderdale.'

Doug left the office of Brigadier Tetley in a daze. He went into the workshops and found Reg Whatley inspecting a tank tread.

Reg was cheerful. 'Ah-ha! The mountain has come to Mohammed. You look awful! What happened?'

'Sit down, Reg.'

Doug explained the previous night's fight over the tank. By the time he was finished, Reg was ashen.

'What a damn fool I've been,' he said.

'It could have happened to any of us.'

'She must have a Vichy contact or someone in the Consulate who tipped her off I had an appointment. I'll come back and make amends.'

'How do you intend to do that?'

'I'll see her again. Get the truth out of her...'

Doug thought for a moment. 'Well, there's just a chance ... Can you come back with us to the coast now?'

'Sure.'

At sunset that day, another wadi was found reasonably close to the harbour and the Tiger was driven there when darkness fell. Once the camouflage nets had been secured, Doug instructed Pumfrey to get hold of some palm fronds and smooth out the wheel tracks. He was taking no chances.

'When any one of us comes here or leaves, we must disguise the imprints of our footsteps in the sand.'

After their recent experiences, none of them thought this was taking precautions too far.

Later that evening, Reggie, a shy man by nature, was still in the dumps. Nothing Doug could say would cheer him up. He knew he had been duped and he felt an idiot for falling for the charms of such an obvious beguiler.

Doug tried to take the edge off Reg's self-consciousness. 'Your girl, Connie, cannot possibly know yet what took place here last night.'

Reg sniffed dismissively. 'Don't try to comfort me. Of course she does. She set the whole thing up.'

'But she won't know that her German officer didn't manage to steal the tank. She'll be awaiting the outcome. You can meet her and give her the news.'

'If I see her again, I'll strangle her.'

'No, you won't. You'll be very upset and behave like a puppy in love. You will give her the impression that you're such a thick clot that you haven't connected her with the incident at all. And, if you play your cards right, you can get out of her who her contacts are and what her Plan B is. How would you normally get in touch with her?'

'I'd phone her at the Consulate.' From his pocket, Reg took out the calling card she had given him.

Doug took it from him and compared it with the card Captain Blanchard had handed over. The cards were identical except for the telephone numbers.

Doug made his decision. 'Ring her now.'

It took another coffee and down-to-earth reasoning to persuade him, but within the hour Reg had phoned Connie and arranged to meet her.

The lounge of the Hotel Seybouse had four guests in it. Two American naval officers in their smart whites were sitting in one corner talking earnestly over a couple of Coca Colas. Dudley Wrangel Craker was sitting on a stool at the bar and appeared to be totally absorbed practising magic tricks with a deck of cards while knocking back a cocktail. Sitting with her back to the room outside on the veranda, and sipping iced water, was the woman who called herself Connie Blanchard.

Lieutenant Reg Whatley entered alone. He took a quick look round the room and went directly across to sit next to Connie. She favoured him with one of her tinkly laughs. 'You look so handsome in your dress uniform. Would you like a drink?'

He shook his head and looked out on to the garden with its palm trees and imitation waterfalls, faintly lit by moonlight.

Connie's face became serious. 'Something is troubling you, Reggie. What has happened?'

'Can't you guess?'

She shrugged. 'Have you been recalled to England?'

Reg looked her squarely in the eyes. 'We nearly lost our tank.'

Connie sat up a fraction straighter. 'Nearly?' she repeated.

'Somebody tried to steal it in the middle of the night. He didn't get far.'

Connie stared at him before taking a sip of water. 'What happened to the man?' she said.

'Got clean away. Gosh, I wish we'd managed to arrest him.'

Connie's mind seemed to be thinking things out. 'So the tank is safe?'

'Absolutely. It's still in the hangar; great place to hide it. I wonder who that man was.'

'Whoever it was, he wouldn't have been able to drive it, would he?'

'Not unless he was a tank man. And why would anyone try to steal it if he couldn't drive it?'

'Well, the important thing is that it's still safely in the hangar.'

Reg forced a smile of apparently dogged devotion. 'I was hoping we might be able to go for another quiet walk together.'

Connie was staring into her glass of water seemingly preoccupied. 'Not today, Reggie, I'm sorry. Sunday is always such a busy day.'

'Is it?'

'When my husband is away, yes. It is left to me to do everything.'

'I suppose there's heaps to do in the Consulate.'

'Exactly.'

'Can't be helped. I'll walk you back there.'

'No! No, please don't. The Consulate is closed now. I have to see someone else.'

'Oh?'

'Yes, I have an important appointment in another part of town. Oh, look at the time. I'm already late.'

Ali, the bar steward, materialised and addressed Reg. 'Lieutenant Whatley, sir?'

'Yes?'

'A telephone call for you, sir.'

'Here in the hotel?'

'In the lobby, sir.'

Reg looked bewildered. 'Who the devil knows I'm here?'

'There are spies everywhere,' said Connie with wide, innocent eyes. 'You'd better take it.'

Reg gave Connie a long questioning look before excusing himself. He left the room and noted that the American naval officers had already left. But Dudley Wrangel Craker was still amusing himself with his cards at the bar.

In the lobby, Doug and Pumfrey were waiting.

'Was it you pretending I was wanted on the phone?'

'Yes. What did she say?' asked Doug.

'She's ditching me. Says she's got to meet someone.'

'Good. Let her go. Don't do anything to make her think you suspect anything. I'll tail her,' said Doug.

'I'll come with you,' said Pumfrey.

'No,' said Doug firmly. 'She's already seen you. She may recognise you, but me she doesn't know from Adam. Go back and say goodbye to her, Reg.'

Reg swallowed his pride and returned to the lounge. Connie's chair on the veranda was empty and there was no sign of her in the garden. He looked over to where Dudley Wrangel Craker had been seated, but his stool was empty too. All that remained was a pack of playing cards. Reg found himself alone with the barman.

'Can I get monsieur a drink?'

'No, thank you. Did you see where the lady went to?'

'No, sir.'

'What about the man at the bar. Where did he go to? He didn't come out past me.'

'I'm not sure, sir,' said the barman, smiling thinly, almost gloatingly. 'Perhaps he and the lady left together. When you were taking your call.'

Reg dashed out to where Doug and Pumfrey were still waiting.

'She's gone,' he told them. 'Must have slipped out through the garden. So has the bloke at the bar.'

'What bloke?'

'I don't know who he was. Chap with a Panama. He's gone now.'

'That was Craker,' said Doug. 'I saw him when I looked in a few minutes ago. Very odd.' Doug knew now who Craker really was, but he was bound by his oath of allegiance to say nothing. 'We meet two strange characters in Bone and they somehow manage to disappear at the same moment.'

'You think he's somehow mixed up in this?'

'I don't know. But I wouldn't be surprised if he turned out to be interested in your Connie. She's just his cup of tea. Probably capable of telling him all sorts of secrets. All we can do is to keep trying her number in the hope she answers. Otherwise, we'll just have to put it down to experience.'

CHAPTER TWENTY-NINE
19 AUGUST 1943

Three days passed without further incident. Reg tormented himself by calling Connie's number several times a day. There was never a reply.

'Probably out of the country by now,' said Doug. 'Or selling her secrets to our other mysterious friend. No matter. The Tiger's safe and no real harm done.'

The new CO at the tank depot had convinced himself that Doug's team was really an undercover commando gang, so he made life easier for himself by leaving the Major to his own devices. This was just as well for only Doug, Pumfrey, Baker and Wilkes knew exactly where Tiger 131 was hidden, and they took it in turns to keep guard on a strict rotational basis.

They pitched their tents on the sand within sight of the camouflaged tank, but there were no obvious signs that they were guarding anything. Pumfrey had scrounged four mosquito nets from somewhere and, as the days grew hotter, the protection made their alfresco lifestyle a little easier.

Occasionally, Doug would look into the Club Exotique and bash out a Fats Waller tune. 'Honeysuckle Rose' was a great favourite. Apart from that, their routine became humdrum. Reg remained inconsolable and eventually appeared to have confined himself to barracks. Doug missed his company.

Mr Dudley Wrangel Craker turned up at the club one evening and listened appreciatively to Doug's jazz piano. However, they had an unspoken understanding that Connie's disappearance was not a topic for conversation. Neither was their meeting in the Brigadier's office. Only once did Dudley drop a hint that he knew more about what was happening in the Mediterranean than he was letting on. 'Wherever German troops have been stationed in French territories, you find Vichy agents, informers and collaborators are operating inside every government department.'

Doug Lidderdale's instincts were to keep moving. Underneath the disciplined uniform of an army officer, there was a 29-year-old jazz-playing, car-racing enthusiast, and the enforced delay in Bone began to wear him down. There was also the matter of an unfinished honeymoon.

The Allies had the 8th Flotilla of submarines based nearby in Algiers. But there was no sign of the promised convoy to take him and his tank back to Blighty. Six weeks after the July invasion of Sicily, the Axis forces had been driven from the island and the Allies were poised to make their main thrust into Italy. The British, American and Empire forces needed every seaworthy craft in the vicinity to transport men and weapons to the new front. Doug hated the waiting but found his priority rating plunging as the demand for more troops and supplies increased. He feared that at this rate he could still be kicking his heels come Christmas.

Every now and then, a British sub or a freighter arrived from the UK to dock briefly in Bone harbour, and invariably this resulted in a flurry of activity as the mail was distributed. Doug received messages from his old CO in the UK and postcards from his mother – certain key words were always blacked out, which often made her news incomprehensible. But there had been no communication from Kate since he left Tunisia. Doug was deeply worried when he wrote a carefully worded letter to his bride on an army aerogram requesting news of her safety. German air raids over southern England had lessened during 1943, and Doug knew it was unlikely his beloved had come to harm. But the horror of such a possibility was never far from his mind.

Doug was lying in his tent, marking up his journal, when he received the news that Reg was leaving North Africa.

'I don't want you to go,' he said frankly.

Reg shrugged. 'My job with you is over. I'm being transferred to another unit.'

'But I need a man with your abilities. You can help me write up the technical bumf that'll be needed once we get the Tiger back to Chobham Green.'

'When a Churchill gets knocked out, they need people like me to pull it out and put it back together again.'

Doug was sure Reg was troubled about something. He studied the Lieutenant's face. Then he said quietly, 'This isn't anything to do with Connie, is it?'

'Good Lord, no, sir.'

'Call me Doug, for God's sake. I know you've got a girl back home, so why don't you take this opportunity to come back with me to see her?'

'I'm under orders, Doug. Big troop movements are afoot. It must be something to do with the push to take back the continent.'

'You really are under orders?'

'Yes, sir. We ship out tonight.'

'I'm going to bloody miss you, Reg.'

'Me too, sir.'

'Doug!'

'Douglas.'

Doug grinned. 'That's what my wife calls me.'

'She's a lucky woman, sir. It's been a great pleasure to work with you.'

'Get out of here.'

Reg grinned widely. 'Yes, sir.'

'And Reggie...'

'Sir?'

'You're already up for one medal.'

'Yes, sir.'

'One is enough. Don't do anything rash. Good luck.'

'And to you, Doug. See you back in Blighty.'

Reginald Whatley opened the flap of the tent and went out into the baking sun.

Doug went back to his journal. 'We'd never have done it without my old pal, Reg,' he wrote.

On the day that Doug posted his anxious letter, Kathleen was in a large army transit camp in Hampshire talking to the Adjutant of the Garrison. Captain Clark was a mild-mannered man of 35 whose thick-lensed spectacles betrayed the reason he was doomed to a desk job for the duration of the war. Poor as his sight was, he could see that Mrs Lidderdale was not a happy woman.

She threw a scruffy envelope on to his blotting pad. 'This has been returned to me,' she said.

The Adjutant noted that the envelope was addressed to Major

A.D. Lidderdale and that it had the appropriate Army Post Box number written on it. The envelope had been neatly opened and on the back in bold writing was the instruction: 'Contents Inadmissible – Return to Sender'.

'What is the meaning of this?' said Kathleen angrily.

'I understand your husband is attached to a tank unit, is that correct?'

'Yes.'

'And he's abroad?'

'I'm not sure that I'm allowed to tell you. In fact, I don't know much myself. I listen to the news on the Home Service to try to make sense of what's going on but it all seems such a shambles.'

The Captain took off his spectacles and rubbed his eyes. Here was a situation with which he was all too familiar. 'No doubt you refer to something in your letter that could be interpreted by the enemy in such a way as to give some suggestion of the nature of the work on which your husband is involved.'

Kathleen's eyes widened. Never slow on the uptake, she responded, 'Were you by any chance a lawyer before you joined the army?'

Captain Clark replaced his glasses and looked at her with surprise. 'As a matter of fact, I was a solicitor. How did you know?'

'You speak like a lawyer,' said Kathleen drily. 'Without any punctuation.'

Captain Clark looked chastened as the myopic eyes behind the steel-rimmed spectacles blinked. 'I stand reprimanded,' he said. 'The fact is, you must have written something which could be of benefit to the enemy.'

'Nonsense. I haven't mentioned his work at all. I don't even know his whereabouts. He could be in Italy or North Africa or...'

Captain Clark put a finger to his lips.

Kathleen was not to be cowed. 'I've only told him about me. How I've teamed up with one of my sisters, Marjorie, and dusted off our old song and dance routine. We fill in when ENSA finds itself with an empty spot. Nothing serious; just to keep our boys entertained.'

Captain Clark's eyes lit up. 'It is you! Of course! I saw you last night at the Garrison Theatre. In the programme it says you used to be one of C.B. Cochrane's Young Ladies.'

'That's all in the past.'

The Adjutant was in raptures. 'Oh, you were marvellous.'

Kathleen was somewhat mollified. 'Well, thank you.'

'Your rendition of "Everybody's Doin' It" was simply wonderful.'

'Didn't you like "Lovely Weekend"?'

'Oh, yes, that was lovely too. Could I make so bold as to ask you to honour me with your autograph?'

'Yes, of course.'

Kathleen leaned forward to sign the proffered piece of paper. Once she handed it back to him, Captain Clark had a question. 'In this letter to your husband, did you mention that you are performing here in Aldershot?'

'Yes. And then in Catterick and Biggin Hill and Oswestry. There may even be a chance of us going to Italy soon. I thought he and I might be able to somehow meet up.'

The Adjutant studied her signature and carefully pocketed it, saying, 'Did you write all this in the letter that the censor rejected?'

'Yes.'

'But, Mrs Crane...'

'Lidderdale.'

'Mrs Lidderdale – sorry – you can't expect to list all these army garrisons and hope that the censor will let them pass. As for Italy, we haven't retaken it yet, so writing of the possibility of giving a concert party there would be frightfully informative to the Italians and the Germans. You must be able to see that?'

The light suddenly dawned for Kathleen. 'Oh dear, one has to be ever so careful.'

'Ever so,' agreed the Captain.

'Thank you for your time.'

'Thank you for your autograph.'

'Goodbye, Captain.'

'Never say goodbye; only *au revoir*. I'm coming to see you perform again tonight.'

CHAPTER THIRTY
SEPTEMBER 1943

The fourth anniversary of the day Britain and France declared war on Germany had come and gone and with it came the belief that the tide in the Mediterranean theatre had turned in favour of the Allies.

Of the 52 U-boats Germany had successfully infiltrated into the Mediterranean through the Straits of Gibraltar, all but 13 had been sunk. This was encouraging news and Doug hoped it might in some way influence his departure from Algeria.

Each day he visited military headquarters and the docks hoping for news of his passage home – or at least a letter from Kate.

One day, Dudley – in his Panama hat and white suit – stood shoulder to shoulder with Doug – in tropical khaki uniform – as they waited on the dockside for one of the scheduled convoys to arrive. Behind them were the mountains, the romantic backdrop to Bone, with green fields and orange groves on the lower slopes.

A mile and a half away, surrounded by curious camels, Wilkes and Baker were guarding the camouflaged tank.

Dudley lit a cheroot. 'Good news, I think. General Jeschonnek has committed suicide.'

'Wasn't he one of Goering's bigwigs?'

'Yes. He was Chief of the Luftwaffe General Staff.'

'So why has he topped himself?'

'One can only surmise that the morale of the German Air Force is at an all-time low.'

'I don't suppose it will affect Hitler,' said Doug. 'He seems to put all his faith in ground artillery and tanks.'

'That's right. And I hear that Herr Speer is developing a new super tank for him.'

Doug suddenly took an interest. 'Do you mean they are upgrading the Tiger?'

'Oh, I'm sure they are refining the Tiger too. But they've designed a prototype of a completely new tank – the Mouse.'

'Oh! The Bosch have come up with a tiny tank this time, have they?'

'I am of the opinion that they christened it in a gay, sardonic frame of mind. My informants tell me it weighs about 200 tons and is armed with an enormous 150mm cannon.'

Doug burst into laughter, which he found difficult to control.

Dudley looked put out. 'What's so funny?'

Doug calmed himself. 'Do you know how many miles the Tiger does to the litre?'

'No.'

'I can tell you that. I'm writing the specification manual. It takes 50 litres of fuel to travel 10 miles. So for a 200-ton tank it would take ... why, they'd have to carry so much fuel there wouldn't be any room left for the driver!'

He giggled again at the absurdity of it, then calmed himself and continued to gaze across the harbour. 'Still no sign of our bloody ship.'

Dudley sensed what was really on his mind. He put a hand on Doug's shoulder. 'Have you not heard from your wife, Major?'

'Not for over a month.'

Doug tried to move away from Dudley's hand.

There was a pause. 'Well, I'm sure she's all right.'

'How can you be so bloody sure?' Doug snapped. 'Have you got spies in England too?'

'Bad news travels fast. No news is good news.'

Doug flicked his cigarette butt into the filthy water of the docks. 'That's a great comfort! Come on. I'll buy us both a drink.'

'At the hotel?'

Doug shook his head. 'No, let's meet at the club. I've lost my appetite for that particular hotel.'

At the Garrison Theatre in Oswestry, Kathleen was on stage contributing to an ENSA concert entertaining conscripts who had just completed their basic training. She was singing 'Stormy Weather.'

> Can't go on –
> Everything I had is gone,
> Stormy weather!
> Since my man and I
> Ain't together,
> Keeps raining all the time –
> Keeps raining all the time...

As she acknowledged the applause, there were tears in her eyes.

CHAPTER THIRTY-ONE
4 SEPTEMBER 1943

It was not until the evening of Wednesday, 4 September 1943 that Doug began to glimpse a chink of light at the end of the tunnel.

Dudley Wrangel Craker turned up unexpectedly at the Club Exotique wearing his customary white suit. He always seemed to have an unlimited supply of cash. He ordered a round of drinks, and settled down to speak.

'For the ancient Romans,' he began, 'this town was a luxurious resort, like Capri. The old name for Bone was Hippo.'

Pumfrey lifted his glass. 'Were there hippopotamuses here in those days?'

Dudley regarded the NCO with a severely jaundiced eye. 'I think you'll find that Hippo is Greek for Horse.'

Pumfrey nodded sagely. 'They don't teach us much Greek in tank school.'

Doug joined them at the bar and sipped the scotch that Dudley had ordered for him. 'So far so good,' he said. 'We may have retaken Sicily, but, if we don't go into Italy soon, I'm going to have to set up home in that bloody tank.'

Dudley put his Panama hat on the bar. 'The invasion started yesterday,' he said casually.

Doug checked himself and stared at Dudley for a few moments. There was one of those lulls in the conversation which some wag has put down to an angel passing.

'I do hope Reg Whatley's safe,' said Doug. Then he downed his drink in one and beamed. 'I think this may call for champagne.'

Doug soon learned that, once the Allies had successfully established an invasion force in Sicily, the decision was made to invade Italy from the coast. General Montgomery, commander of the 8th Army, stood up on the seat of his Jeep to address his troops.

'We are going to step on the gas and get into Italy,' announced the General. 'We can see the end. I don't say that it will be this week or this year, but we can see daylight.'

The Club Exotique occasionally supplied what it described as champagne. It was certainly fizzy and the false label pronounced it as being of very high quality. But its provenance, let alone its vintage, was dubious in the extreme. Nonetheless, a week later, when it was announced that Marshall Pietro Badoglio had surrendered the Italian army to Allied forces, they ordered a bottle. The cork bounced off the ceiling just like the proper stuff.

'This is freeing up the Med,' said Doug. 'I've had my fill of the so-called delights of bivouacking under the stars.' He went on with uncharacteristic bluntness. 'When is this blasted boat coming to pick us up?'

He repeated that they had been waiting since Wednesday, 11 August.

Dudley waited patiently for Doug to vent his frustration before pricking the bubble. 'A convoy has been convened. It's due to leave Alexandria on 14 September and should get to

Bone harbour six days later. It's a big, slow convoy, and it will be stopping at nearly every port on the way to Gibraltar. But this might well be the one you've been waiting for.'

'It'll be here by the 20th, you say?'

'That is only an estimated time of arrival.'

'Thank God for that; a definite date at last.'

'That all depends.'

'On what?'

'On the minesweepers being satisfied they've cleared the coast of Rommel's asparagus.'

Doug's face registered total non-comprehension. 'What the devil are you talking about?'

Dudley explained. 'Before Rommel was forced to retreat, he was determined to make it impossible for the Allies to get supplies into the seaports along the North African coast. So these waters are treacherous, full of underwater hazards such as jagged rocks, sunken concrete slabs and timber ramps and, of course, God knows how many floating mines just beneath the surface. The sweepers are still in the process of clearing them. There are so many things sticking up under the water they've become known as Rommel's asparagus. Even when you get into the open sea, you will only be relatively safe as far as Gibraltar.'

'Well, that's a thousand miles closer to home,' said Doug.

'Yes, but once you've passed Portugal you have to go through the Bay of Biscay. It's along this coast that Admiral Doenitz keeps his U-boats. Doenitz himself is based at Lorient, but he has bases in Bordeaux, La Rochelle, St Nazaire and Brest. I'm afraid that's where you're going to be running the gauntlet.'

'How many subs has he got?'

'They are churning them out on a daily basis. Several hundred.'

Doug filled their glasses with bubbly. 'I wonder if it would be possible to make a raft for our tank,' he mused.

Pumfrey scoffed. 'To carry 60 tons? I don't think so.'

'Winston Churchill wants this tank. And, by God, I'm going to deliver it to him. If we get hit by submarines, we must have some sort of safety net to save the Tiger.'

Dudley turned to the surly Frenchman behind the bar. *'M'sieur le patron, s'il vous plait, un autre Dom Perignon Superb, merci.'* He turned back to Doug. 'You know, this may be a three-bottle problem, as Sherlock Holmes might say.'

U-boat commanders told Admiral Doenitz that the Allies were listening to them but he dismissed this information, saying, 'Impossible, the code machines cannot be broken.'

The irony of this was that the British Admiralty's code had been broken by B-Dienst, the German intelligence service who fed convoy details to Doenitz. Incredibly, the Germans never recognised any of their own Enigma information contained within the Admiralty signals.

In June 1943, German suspicions had been sufficiently awakened for them to reluctantly change their entire communications network to a new cipher system known as 'Medusa', or 'Turtle' as the Bletchley Park scientists were to name it. 'Medusa' was even more of a mental ballbuster. The cipher was so fiendish that the German mathematicians who devised it were confident that, this time, it really was unbreakable. They underestimated the crossword addicts in Hut 8. The new code was cracked within 24 hours. This added breakthrough gave the Allied convoys a sporting chance of outmanoeuvring the U-boats.

CHAPTER THIRTY-TWO
17 SEPTEMBER 1943

The mysterious Dudley Wrangel Craker vanished for long periods. On one of the occasions he rematerialised, he sat listening as Doug picked out the melody of 'Stormy Weather' on the piano at the Club Exotique. Doug addressed Dudley as he played.

'Did you glean anything more about that Jerry who tried to steal our Tiger?'

'Not much. You probably know that Tiger 131 had belonged to the German platoon officer – Number Three Platoon, actually.'

'Yes, 504 Heavy Tank Battalion. I found that out myself.'

'Well, the curious thing is, since Tiger 131 was abandoned, none of her crew has ever been found.'

Doug affected surprise. 'Is that so?' he said, sipping his drink.

Dudley stroked his broken ear. 'If I didn't know better, I'd say it was the work of a commando raid.'

Doug stared Dudley straight in the eye. 'Something tells me that you do know better.'

Dudley shrugged. 'Number Six Commandos were not in the

area at the time. So, the fact is, nobody knows whether the tank crew got killed or whether they ran away.'

'Is that right?' said Doug, sounding noncommittal.

'I don't suppose anybody will ever find out.'

'Probably not. So who was the man we caught?'

'A Major attached to the second SS Panzer Division, Das Reich. The equivalent rank to yourself, Major Lidderdale. If his identification number is correct, he is someone called Hans Kruger. He is an SS man, an Abwehr spy. Very blonde, very proud; very difficult to break.'

'I take it that somebody did try to break him?'

Dudley shrugged again. 'I wouldn't know, old boy. Things come to me on the grapevine. I'm only a cultural adviser for the British Council.'

'Of course, Dudley; so you told us.'

'Anyway, Major Hans Kruger escaped.'

'Good God! How did that happen?'

'He was allowed to. The idea was that he would lead us back to whoever was controlling him.'

'And?'

'We lost him. He disappeared.'

'So he's on the loose?'

'Yes.'

'Is he likely to have another go?'

'A rhetorical question, Major. We know who he is. We know what he looks like, so he's unlikely to hang around. It wouldn't surprise me if he isn't back in Germany by now.'

Doug looked uncomfortable. 'I hope to God you're right.'

Dudley continued in thoughtful mood. 'Interesting thing about junior SS officers ... were you aware that Himmler will not allow them to marry? He doesn't want them to taint their Teutonic stock with outside ethnic blood.'

'Good Lord! Are they all virgins then?' asked Doug.

'By no means. They're encouraged to father bastards of Aryan purity. But, if there's a shortage of Teutonic ladies, there are special brothels provided: one for officers, another for other ranks.' Dudley smiled and added, 'The Germans are renowned for their efficiency.'

'Don't think I'm rude, but what did you do before the war?' asked Doug.

Dudley's face took on a wistful look. 'I was always organising.'

'Organising what, exactly?'

'Oh, things like the Royal Tournament. Do you remember? 1925?'

'I didn't see it.'

'I organised that. We had a pageant portraying imperial artillery through the ages. About 50 field guns pulled by elephants and 14 of the biggest Nigerians I could find.'

At that moment, Corporal Baker ran into the bar. 'Major!' he shouted. 'Major Lidderdale – come quickly, sir. The convoy's arrived.'

The Ministry of War Transport Department had a warren of offices at the docks of Bone and it was here that Major Douglas Lidderdale and his small crew convened. The CO for the Transport Ministry was a harassed Colonel with far too much on his plate. He introduced Doug to Captain William Rickard, Master of the SS *Ocean Strength*, and then excused himself from the room.

The Captain was not old, but he looked an ancient mariner. The hazards of commanding a transport ship in wartime had started to take their toll. However, after shaking hands with everybody, he rallied and came straight to the point.

'I'm told you want me to transport a rather special tank back to England?'

'That's right, Captain,' said Doug. 'We've been waiting for you since the beginning of August.'

Rickard sat down and wiped a weary hand across his forehead. 'Sorry about that. We invaded Sicily, you know. And, if that wasn't enough, we then took on Italy. The Italian fleet has escaped, but our troops seem to have landed their boots on pretty solid terra firma.'

'Yes. We've had our ears glued to the BBC World Service.'

'*Ocean Strength* has been a supply ship on both operations. I couldn't get away any sooner. Sorry about that.'

Doug laughed. 'Well, I'm glad you got round to picking us up at last.'

Captain Rickard took out a dirty-looking pipe and started applying the flame of a windproof lighter to it. 'Let's hope that we'll be able to weigh anchor soon enough. We know of a rogue U-boat, UJ-2104, in the vicinity, and I'm afraid that might cause a delay if we can't pin it down. By the way, have you heard that hundreds of German paratroopers dropped into Rome and rescued Mussolini?'

'When did this happen?'

'Last week. Perhaps it hasn't filtered through to the BBC yet. But, if Mussolini gets back on his feet, we could be in trouble.'

'In that case, let's get under way as soon as possible.'

'I've got to refuel. And it's Saturday, so they won't do it today.'

'Who won't?'

'The locals. They're Mohammadens. Won't lift a finger on Saturday.'

'Tomorrow, perhaps?'

Rickard sniffed and sucked on his evil-smelling pipe. 'Sunday. Knackered. Christians.'

'Oh.' Doug sighed. 'What about Monday? Or is that when the Plymouth Brethren take the day off?'

Rickard withdrew his pipe and looked vaguely optimistic. 'Monday sounds good. Where are you going to put your tank?'

'On the deck, I suppose. Can't possibly get her into a hold.'

'You could lash her to the foredeck.'

'That sounds promising.'

'Big tank, is she?'

'Over 60 tons, plus another 10 tons in spares.'

Rickard dropped his pipe to the floor. 'Good God!' he said.

Corporal Pumfrey put a word in. 'That's funny. Everybody says that the first time.'

That night the Club Exotique came close to living up to its name. The crews of the ships in port were on shore leave for the first time in weeks. The ladies of the night had had a lean time of it lately, but they were all set to make a bob or two tonight.

Captain Rickard introduced his Chief Officer James Henderson and his Chief Engineer William Bain. To the mutual satisfaction of all concerned, it was discovered that they each had a thirst on them that would not be satisfied until they felt their teeth float.

Pumfrey pointed out that 'Our Man in Cairo' – Dudley Wrangel Craker – was nowhere to be found.

Wilkes had been left on guard duty with the Tiger, so Pumfrey departed from the party at 0200 hours to relieve him. This turned out to be fortuitous because Pumfrey was the only one without a headache the following night, a Sunday, when Tiger 131 had to be driven under cover of darkness on to the foredeck of *Ocean Strength*. The tank's enormous weight meant this was not a straightforward proposition.

The ramps from the quayside to the deck of the ship had to

be reinforced with rail tracks. Doug watched as the Chief supervised the sailors as they lashed the Tiger to the deck with thick ropes. He chose to use his own team of men to conceal Tiger 131 under camouflage netting to distract any prying eyes during the graveyard shift.

Once all was secure, he went to the bridge where he shared a weak but welcome coffee with Captain Rickard, who read a dispatch bringing welcome news. The Royal Naval destroyers *Echo* and *Intrepid* had sunk U-boat UJ-2104.

CHAPTER THIRTY-THREE
20 SEPTEMBER 1943

S S *Ocean Strength* looked somewhat unimposing painted in wartime grey. Before the ship left her moorings at Bone on Monday, 20 September 1943, Dudley Wrangel Craker – representative of the British Council to some, intelligence officer to others, war reporter for *The Times* to the rest – arrived with his hat and bow tie to bid bon voyage to his newfound friends. He shook hands with each of them and ended by looking shrewdly into the cool grey-blue eyes of Major Douglas Lidderdale.

'Good luck, old chap. Don't let your guard down for a minute. Until you get back to the British Isles, you'll be in grave danger. The Jerrys want their Tiger back. And, if they can't get their hands on it, they would prefer to send it to the bottom of the sea than let it become some sort of trophy in London. Trust me, I know of what I speak. I have my sources.'

'Escort ships will be screening us, Dudley.'

'My dear Douglas, don't you think the Germans know that? Admiral Doenitz and Field Marshal Goering will be following your every movement. Whether they have the capability any longer to attack you in the Mediterranean is a moot point, but, as soon as you get past the Rock and into the Bay of Biscay, there will be whole packs of submarines and flocks of Focke-Wulfs after you.'

'You're very encouraging.'

'I'm concerned for you. I don't want you to underestimate the threat.'

'Thank you, Dudley. I know you've done a lot more for us behind the scenes than you can talk about.'

'Nonsense, dear boy, I'm merely a cultural adviser. By the way, take this.' He handed Doug a package the size of a shoebox. 'You never know when you might need it.'

'What is it?'

Dudley lowered his voice. 'A knife – a dagger to be precise. It's called the Fairburn-Sykes fighting knife. Blade is seven inches long. Very good for sentry annihilation.'

Doug frowned and stared uneasily at the box. 'I've got my service revolver.'

Dudley graced Doug with a charming smile. 'I know. But there may be times when you may need to dispose of someone without making one hell of a noise.'

Before Doug could protest, Dudley continued. 'One more thing. You'll find a large number of inflatable life rafts in the number one hold.' He said for Pumfrey's benefit, 'That's the goods store underneath the front deck. There are five holds in this ship. Yours is number one, at the front end.'

'Life rafts, eh?' said Pumfrey. 'You don't set our chances of a clear run too highly.'

Dudley looked slightly offended. 'But you asked for them.'

Doug frowned. 'We did?'

'Yes. You wondered if the Tiger could be made to float if the worst came to the worst.'

Pumfrey laughed. 'I hardly think that 60-odd tons of dead weight could be carried on a life raft.'

'There are 60 life rafts,' said Dudley. 'Commandos use them all the time.'

Doug and Pumfrey stared open-mouthed.

Dudley went on. 'One raft per ton. It's entirely up to you how you attach them to the Tiger.'

Dudley's warning was to prove prophetic. On the night SS *Ocean Strength* edged out of Bone harbour, the Captain of U-boat U-221 received a coded signal from Admiral Doenitz's headquarters in Lorient. The U-boat was instructed to detach itself from the Leuthen patrol, the Wolfpack operation formed to hunt down ON-ONS convoys in the North East Atlantic.

'You are to become a lone wolf,' read the message. 'Convoy MKS 025 is sailing from Algeria 20-09-43 bound for the UK. Seek and destroy. Priority. Doenitz.'

Kapitanleutnant Hans-Hartwig Trojer immediately assembled his crew and made preparations for U-221 to cast off from its base in St Nazaire. U-221 had been scheduled to sail with the rest of the 6th Boat Flotilla, but it had been delayed because there had been a temporary hitch while fitting the new acoustic torpedo system, the T-5 Zaunkonig.

At the age of 27, Kapitanleutnant Hans-Hartwig Trojer had been decorated for valour and achievement. He had received the Iron Cross – both 1st and 2nd Class – and the Knights Cross. In his first four patrols, he had sunk 21 Allied ships amounting to the loss of over 70,000 tons. Despite these triumphs, at the start of his fifth patrol, the Kapitanleutnant had spent less than 192 days at sea.

Hans-Hartwig was one of Doenitz's protégé commanders. He was born in Siebenburgen in Transylvania, and for this reason his crew of 50 men nicknamed him 'Count Dracula'. This was not altogether an affectionate soubriquet. Although women thought of him as handsome with his sleek black hair, bedroom eyes and chiselled chin, and men saw in his demeanour the mischievousness of a daredevil, there was something of the night about him.

As he sailed into the Bay of Biscay before setting a leisurely course southward, Hans-Hartwig had every reason to believe that this new mission would have a successful outcome. His submarine had been re-equipped with the latest 38mm guns and the newest radar – FuMB 8 and FuMB 30 GEMA – and all he needed was a day or two to test out his new equipment. He sat in his cramped cabin and made a few calculations. At convoy speed, he estimated that *Ocean Strength* and its precious cargo would pass along the Portuguese coast at some time between 23 and 25 September. His handsome face broke into a crooked smile of self-satisfaction. U-221 would be lying in wait for the Tiger.

Ocean Strength had sailed from Sicily to Italy to Malta and on to Alexandria where she formed part of the huge Mediterranean Convoy officially designated MKS 025. The code was Admiralty shorthand indicating M for Mediterranean, K for United Kingdom and S for slow, and it happened to be classified as a military convoy.

In the misty light of dawn, on their second day aboard, Doug Lidderdale stood on the bridge staring out to sea and was a touch awestruck as he surveyed what appeared to be not so much a convoy as an Armada. There were so many ships distributed over such a wide area that it was impossible to count them all. Doug thought the scene was reminiscent of a Turner painting.

234

The majority of the ships were designated to sail in this convoy only as far as Gibraltar, after which they would continue to various destinations such as New York or Augusta. Of the 71 merchant ships that surrounded *Ocean Strength*, 31 would stay together past the Rock.

The sheer scale of the exercise and a number of damaged stragglers meant a convoy speed of 11 knots could not be achieved. Therefore, they limped along at a less than spectacular rate. Captain Rickard, looking weary and ashen, his spectacles askew, came out of the radio room holding the latest report. His hand was visibly trembling.

'What news, Captain?' said Doug with concern.

'The Germans have started using acoustic homing torpedoes.'

He looked down at his piece of paper and read aloud in a flat voice: 'In the past two days, U-boats in the North Atlantic have sunk HMS *Lagan*, HMCS *St Croix*, HMS *Polyanthis*, SS *Frederick Douglass*, SS *Theodore Dwight Weld*, HMS *Itchen*, SS *Oregon Express*, SS *Skjelberg*, SS *Steel Voyager* and SS *Fort Jemseg*. SS *James Gordon Bennett* is still afloat but crippled.'

Doug was appalled as he listened to this catalogue of destruction. There was a moment's reflective silence before he spoke quietly. 'And we are next to sail into the hornet's nest?'

Captain Rickard nodded grimly. 'We'll use zigzag tactics, of course; and we'll try to go round them. We have sonar and the latest Asdic machines...'

He stopped speaking and suddenly appeared very old and very tired. He put forward his hand to grab a rail, but missed it. His foul-smelling pipe dropped to the deck and the stem broke.

Doug went to retrieve it but the Skipper told him to leave it. Doug took hold of the older man's arm and steadied him.

Captain Rickard spoke so softly he could barely be heard.

'I've been in this game too long,' he whispered. 'I was going to show you round the ship, but please excuse me if you don't mind – I think I'll go to my cabin for a while.'

'Of course,' said Doug. 'Can I help you?'

'I'd rather be left alone. I had friends on those ships.'

There was a silence. It was a difficult moment. The old man was seriously upset. The loss of so many ships in the 48 hours since they had left port was a cruel reminder to the tank specialist of the danger he faced and of the fact that there was no turning back now.

Captain Rickard recovered his composure. 'I'll ask my Chief Officer to do the honours,' he said softly as he walked away.

Chief Officer James Henderson joined Doug on deck after a lunch of Spam and reconstituted potato.

'The Captain asked me to look after you. He is not feeling so well.'

'Yes,' said Doug. 'He seemed like a broken man. Will he be all right?'

'He needs rest. He'll be retiring at the end of this voyage. Now, let me conduct you on the scenic tour. Half a crown and ice cream is discretionary.'

Doug and the Chief walked briskly aft, where they found Pumfrey and Wilkes staring in fascination at the dolphins following in the *Ocean Strength*'s wake.

'This ship is 441 feet and 6 inches long,' said Henderson. 'That extra 6 inches is the important bit, as any lady worth her salt will tell you.'

Doug smiled politely.

'We run on fuel oil which is carried in the inner bottom tanks just above the keel of the ship. Here, on the stern deck, you can see a four-inch gun flanked by two smaller guns. The four guns with the long barrels and the long spindly legs that you can just make out on the superstructure are 20mm cannons called Oerlikons; very effective against U-boats. They fire magazines of 60 rounds. And, for good measure, gentlemen, there are machine guns either side of the forward mast. There are four guns fitted adjacent to the second mast and a three-inch gun on the bow, so you can rest assured we can defend ourselves if absolutely necessary.'

'Why are they called "Liberty Ships"?' asked Pumfrey.

'Building them wholesale is an American enterprise, and they are currently being built at the rate of three a day. President Roosevelt says these ships will bring liberty to Europe. The phrase has stuck.'

Doug added, 'Perhaps he was thinking of Patrick Henry's speech of 1775 when he wrote, "Give me liberty or give me death."'

The others stopped walking for a few moments and stared at Doug.

'Forgive him, Chief,' said Pumfrey. 'He is the brain in the family.'

They approached three sailors engrossed in private conversation which stopped the moment they became aware of the Chief and the REME soldiers. One of the sailors was tall and fair-haired. He looked apologetically at the Chief and Doug.

'Just getting some fresh air, Chief,' he said, and started down the hatch.

'Wait a minute, sailor,' said Henderson. 'Where have you come from?'

'Engine room, Chief.' He disappeared below. He had a foreign accent.

The Chief looked pensively at the hatch. 'He must have joined us in Algeria. It's difficult to keep track of the crew.'

'How many are there?' asked Doug.

'Forty-five crew. Thirty-six gunners.'

They approached the foredeck of *Ocean Strength* where the Tiger was lashed down and completely covered in camouflage netting.

'What about our escort ships, Chief. Aren't they destroyers?' asked Pumfrey

'Not all of them. After we have got past the Rock, we will have six ships escorting us. At the moment, we are being led by a minesweeper from Sweden but I don't know if she will come with us all the way to Scotland. The other six escorts will be corvettes.'

'I'm sorry, Chief, but what is a corvette?'

'It's smaller than a frigate and it's too fast for submarines to get much of a fix on it. It's lightly armed and good for anti-aircraft fire 'cause it can dodge about like a flyweight boxer. Our corvettes are *Bergamot*, *Bluebell*, *Bryony*, *Camellia* – and *La Malounie*.'

'The only one not named after a flower.'

'Quite. Traditionally, all our corvettes are named after flowers but *La Malounie* was French. When the Froggies capitulated, the British confiscated her. Another of our escorts is called *Jumna*, an A-class sloop, so we have some lovely ladies screening us, gentlemen.'

Chief Officer Henderson then went on to explain about the ship they happened to be sailing on. 'The so-called "North Sands" ships were built to the original "Empire Liberty" design. Sixty were built; half on the West Coast of America at Richmond, California, and the remaining thirty, including SS *Ocean Strength*, on the East Coast at South Portland, Maine.'

Doug expressed his relief that they were on the high seas at long last. The Chief listened sympathetically as Doug explained their frustration at having to sit around and twiddle their thumbs from 11 August until 20 September until a suitable convoy showed up.

'Well, now that we have shown up perhaps you could explain something to me. What's so important about this blasted tank?' asked Henderson.

'Well, it's twice as big as any tank made before,' said Doug, 'and it's twice as mean.'

Henderson realised Doug was not going to be giving much away. 'It must be pretty powerful then?'

'Let me put it this way, it takes five Churchill tanks to destroy a single Tiger, but only one Churchill will survive the encounter,' said Doug.

'Tell me more. How did you get hold of it?'

Doug smiled and tapped ash off his cigarette. 'I can't say any more, Chief. Under orders, you know.'

The conversation was abandoned abruptly as the warning klaxons sounded. Simultaneously, the decks swarmed with men in tin hats and flak jackets.

The Chief started to run back to the bridge. He shouted over his shoulder, 'Enemy aircraft. Tin hats and life jackets. Now!'

Doug and his team scattered like hounds as the sound of fighter planes screamed out of the sky.

'I think I'll pop downstairs,' said Wilkes, 'and inspect the bully-beef stew.'

CHAPTER THIRTY-FOUR
SEPTEMBER 1943

They were Stukas; dihedral, gull-winged, Junker 87 dive-bombers. Flying in formation from the north at high level and then peeling off in vertical dives, they switched on their screaming sirens as they dived.

Anti-aircraft fire started up from all the escort ships within range but the speed of the Stukas made their chances of getting hit very slim. It seemed that the convoy was at the mercy of the German squadron who seemed able to drop torpedoes at will. There was no direct hit on a ship in the first wave, but then the Junkers soared to 2,000 feet again, circled and began a shallower dive.

Helmeted and wearing a flak jacket, Doug observed how things developed through his Barr and Stroud 7 x 50 binoculars. He shaded his eyes and pointed high into the western sky. Half a dozen tiny dots grew bigger as those on the bridge of *Ocean Strength* watched. The six dots materialised into twin-engined Mosquitoes, British multi-role combat aircraft affectionately nicknamed 'Mossies' by their crews and known as Matrons for their ability to protect convoys.

The de Havilland DH.98 Mosquito was the fastest operational aircraft in the world when it first rolled off the production lines in 1941. The design incorporated a 20mm cannon under its nose. Commanders of Kriegsmarine U-boats – of which significant numbers had been sunk by Mossies – feared it, and rightly so.

The Mossies, following the traditional tactics developed by light planes such as Spitfires, came down to attack the Stukas. The idea was to pull out of a dive from behind the victim and fire up at its underbelly. Their strategy was standard and lacked subtlety but their success rate was phenomenal.

The squadron leader in the leading Mossie pushed his control column forward and zoomed down in a near-vertical dive. The other Mosquitoes followed their leader. The Junkers could not have seen them coming out of the sun for they took no evasive action. Then the leading Mossie pulled out of the dive and came up behind and below the enemy's tail, which he lined up in the red-crossed lines of his sight. Then he thumbed the firing-tit.

The German pilot could not have known what hit him. His machine broke up, first an elevator, and then the entire tail snapped off, narrowly missing the speeding British plane following behind. The Junker slipped into a spin and the German plane plunged uncontrollably into the drink.

Messerschmitt 110s appeared from nowhere. These were fast, twin-engined fighter-bombers with a crew of three. They acted as escorts to heavy bombers and were more than a match for the Mossies. They had probably come across this aerial battle while on a patrol cruise. They tried to hunt in packs called Jagdstaffels. A staffel consisted of 12 planes.

Tracer bullets streamed in every direction. The Mossies had selected a target each but had not been prepared for the Messerschmitts flying in to the German rescue.

The ships of the convoy stopped firing since no gunner wanted to risk shooting down one of his own. They watched as Allied and Axis planes crisscrossed each other in a kaleidoscope of confusing patterns.

At the eleventh hour, the cavalry arrived – three Canadian-crewed Beaufighters sped across the sea towards the melee.

In his Junker, Ofw Kurt Gaebler of KG40 Patrol swung round to meet the latest menace. Using the age-old tactic, he climbed high and then screamed down vertically to slipstream the backmost Beaufighter. Tracer bullets from a Mossie diving towards him from the right streamed like neon strips in his direction. Gaebler's right foot exerted pressure on the rudder bar and swerved from the tracer bullets. The pilot of the Beaufighter sought to save himself and started to climb above the foray, inevitably losing speed as he did so. Gaebler, right behind him again, was so fast he was momentarily in danger of overtaking the Beaufighter.

He held the fighter in the cross-lines of his sights, while closing into an effective range. The rear fuselage of the British plane was dead centre in his sights. He could not miss. He triggered a long burst of cannon fire which raked the Allied aircraft.

Smoke and flame trailed out of the superstructure. A fuel tank exploded. Then the big fighter started to spin clockwise as she lost momentum and spiralled down in a vicious stall. Kurt knew that nobody was likely to have survived that. As he flew out of the smoke, he found a Mosquito racing past him in an attack on one of the Messerschmitts. Kurt pressed the firing mechanism again. He didn't have to bother to sight the Mossie. His cannon raked across the belly of the Mosquito, which disintegrated in mid-air.

Kurt had run out of ammunition. He climbed to his ceiling

height and headed for home; he could claim two more kills for 26 September 1943.

Convoy MKS 025 lumbered into the port of Gibraltar, and later the same day *Ocean Strength* hove to. The ships *Adabelle Lykes*, *Brockholst Livingston*, *Chertsey*, *Empire Sparta* and *Ocean Vagrant* docked. The remaining ships split up and joined other convoys bound for various ports around the world.

The ships *Cape Sable*, *Clan MacNair*, *Inventor*, *James Woodrow*, *Rajput*, *San Francisco*, *Trevorian* and *Ocean Strength* sailed on towards the UK. Twenty-three more ships joined them when the convoy reconfigured at Gibraltar and this smaller and more vulnerable convoy sailed on through treacherous waters to the United Kingdom. Their cargos included palm kernels, frozen meat, iron ore, manganese and TNT.

The Chief reassured his passengers that the seas close to the Portuguese coast were in neutral territory and should prove safe. After that, when they sailed into the North Atlantic adjoining the Bay of Biscay, the real danger would begin.

Pumfrey asked the Chief what the official designation was on the *Ocean Strength*'s manifest.

Henderson grinned and said, 'Iron ore.'

'I suppose that's one way to describe a tank,' said Pumfrey.

'It would be a very apt description if we don't get home with what we are carrying in numbers four and five holds.'

Pumfrey raised a questioning eyebrow.

The Chief whispered, 'TNT. Don't spread it around.'

'Is there much?' asked Pumfrey.

'Three thousand tons.'

Pumfrey looked puzzled. 'If we get holed or something and water gets in, that'll be a real waste.'

'Oddly enough, it wouldn't. TNT is one of the few things that is not affected by sea water.'

Only the ships *Fortol*, *Flaminium*, *Fort McMurray*, *Baron Fairlie*, *Ocean Strength* and the rescue vessel, *Copeland*, were due to sail all the way to Glasgow. The Germans guessed that Tiger 131 was on one of these ships but they didn't know which one. Condors regularly circled at 10,000 feet keeping an eye on the convoy.

On Tuesday, 28 September, Doug and his unit found themselves heading for the choppier and more menacing seas of the Bay of Biscay.

CHAPTER THIRTY-FIVE
SEPTEMBER 1943

The convoy steamed slowly northwards. Visible high above in the late-afternoon sun, but out of range, Focke-Wulf Condors 200s – big four-engined planes – circled hungrily like vultures. Nobody could tell whether they were up there purely for reconnaissance or whether they were preparing to drop their 250-kilo bombs. Finally, they disappeared as the last rays of the sun dissolved into blackness. Nobody kidded himself they had gone for good.

Following this flight, III/KG 100, based at Istres near Marseilles, ordered all 18 operational Condors to regroup, shadow and attack the convoy again.

They were armed with the new Fritz Xs carrying Do 217K-2s. The Fritz X was an armour-piercing bomb fitted with wings and could be guided to the target by the navigator who followed its progress visually while operating joystick control devices. Ocean Strength was saved by bad weather. Over the next five days, the Condors only found three ships and no hits were scored.

Every night, the ship's crew resumed normal defence stations. Vigilance in teams – four hours on, four hours off. According to

Standing Orders, lifebelts were always to be worn during action and when sailing in enemy waters. Generally, however, the lifebelts were employed as pillows, for the crew rested between various duties; not at set hours, but when they could.

The Navigating Officer seemed umbilically attached to his binnacle. The pervading sound was the pinging of the ASDIC, the submarine detector, which over time got into the inner souls of the officers of the watch. ASDIC was an acronym for Anti Submarine Division Investigation Committee, or so it was claimed. In fact, it was a deliberate obfuscation. The ASDIC machines contained quartz piezoelectric crystals, and these were part of a top-secret device. Even the operators did not know how it worked.

On the bridge, the only lights were those that illuminated the binnacle, the compass, the charts and the essential phone line.

Doug went on to the deck and looked out across the sea. The escort ships were in darkness. He thought he could see shapes but he knew it was possible his eyes were playing tricks on him. He put a cigarette between his lips, and was considering whether he could risk striking a match to light it when, out of the corner of his eye, he noticed something move on the foredeck. There was no moon at that moment and he found it difficult to focus aided by nothing but the glow of the stars.

But there it was again. At the port edge of the camouflage netting that covered Tiger 131, there was something, or someone, moving.

As far as Doug was aware, nobody else was on deck, not on the port side anyway. Quickly and noiselessly, Doug ran to his cabin and returned with the parting gift given to him by Dudley Wrangel Craker – the commando knife.

Quietly, he approached the foredeck where Tiger 131 lay under the nets. Clouds scudded across the sky, obscuring even the stars. There was no question of hiding in the shadows. Everything was in shadow now. Doug strained his eyes to see if there was still someone lurking by the tank. And then the moon made a guest appearance for a few seconds – just long enough for Doug to see a figure bending over the guy ropes that held the netting taut. It appeared to Doug that the man was hacking at them in an attempt to loosen them.

Doug crept up to the man and, with the dagger against his throat, yanked the man to his feet. 'So we meet again, Herr Jurgens.'

The man started to struggle. 'Get away from me. Help. Stop him somebody.'

It was not the reaction Doug had been expecting.

The Chief Officer ran down from the bridge with a torch on a narrow beam. Two gunners, still wearing flak jackets and helmets, ran to his assistance.

'What the devil is going on?'

'This is a German soldier from a tank unit,' said Doug. 'His name is Major Hans Jurgens of the SS and he was trying to release the camouflage nets.'

The man appeared to be shaking with fear. If it was an act, he deserved an Oscar. He started to cry. 'The rope was loose. I was trying to help. I was putting it back together again.' He had a pronounced accent, but it was not German.

Henderson shone his lamp in the suspect's face. It was the man from the engine room who they had bumped into earlier.

The Chief spoke harshly. 'Just who the hell are you? How did you sneak on to my boat?'

The man cowered and fell to his knees. 'Please. I am Swedish. My name is Sven. I am normally on the minesweeper but I missed the boat. Please. I come on here instead. I have the papers. Please. I am so sorry.'

'You have your certificates and papers?'

'They are in the engine room where I have my friends. They are helping me. I am so sorry.'

The Chief looked at Doug. 'Major Lidderdale, what do you make of this?'

Doug put his knife into its sheath and studied the stranger's face carefully. At length, he spoke. 'I may have made a mistake.'

'Let's sort this out in the map room,' said the Chief. 'Someone go down to the engine room and find his papers.'

Half an hour later, Doug was shaking Sven's hand and apologising. It turned out that the Swede had been with the same convoy of ships for three months. There was no doubt about it. Many of the sailors on board the *Ocean Strength* knew him from their periodic shore leaves.

A little later, Lance Corporal Pumfrey knocked on Doug's cabin door.

'Come in.' Doug was sitting on the edge of his bed looking forlorn.

Pumfrey shut the door behind him and took out a half-bottle of Navy Rum. He went over to the washstand, rinsed the tooth glass and filled it with the brown stuff.

'Where did you get that? There's no drink on these ships. American rules.'

'I heard that before I came aboard, sir. I thought I'd better bring a supply for medicinal purposes.'

He thrust it into Doug's hand. Doug grimaced and took a swig. He gave the glass back to Pumfrey, who drank the remainder.

'It was a very easy mistake to make, sir. Any one of us might have done the same.'

'Thanks, Corporal. But it's my own damn fault. I'm jumpy and jumping to conclusions. The Tiger has taken over my life.'

'I'll bet a hundred to one you'll get a promotion out of it.'

'Oh, I don't care about that.'

'Oh, yes you do, sir. Extra bunce. And you'll need that being a married man.'

Pumfrey poured the rest of the rum into the glass.

'We're on duty. We shouldn't really drink.'

'We're on a ship, sir,' said Pumfrey. 'I think it's part of the rules and regulations to drink your rum ration. Put a spot of water in it and call it grog. You deserve it. Also, sir, there are no teetotal Yanks on board to complain either.'

Kapitan von Forstner was patrolling U-402 at periscope depth from the Atlantic towards the Bay of Biscay. There was a rumour circulating among the senior commanders of the German fleet that Adolf Hitler had personally ordered that the British ship carrying the stolen Tiger tank was to be sunk. Von Forstner had no means of knowing the veracity of the story but he was ambitious enough to try to achieve this small victory.

On Friday, 24 September, U-402 lay in wait in the Atlantic when an order came through which shattered von Forstner's hopes. BdU decreed that U-402 and 18 other U-boats form a new pack called 'Rossbach' and move northwards towards Greenland.

Von Forstner was tempted to ignore the order, or to pretend that he had not received it, but he knew that to disobey Admiral Doenitz in any particular would lead to a court-martial. Reluctantly, he ordered the sub to turn about.

Doug knew nothing of these events occurring around him, nor that luck was still on his side.

CHAPTER THIRTY-SIX
SEPTEMBER 1943

Count Dracula, otherwise known as Kapitanleutnant Hans-Hartwig Trojer of U-221, had been correct in his calculations about the progress of the convoy. Or, rather, he would have been correct had the convoy progressed at the standard speed of 11 knots. However, this was not the case. What is more, the Luftwaffe had suddenly become very active. Goering was no less guilty than Doenitz of opportunism and his planes had been trying to bomb the hell out of the convoy while it was shuffling around Gibraltar, separating into sub-convoys for the various destinations. Fortunately, there were a large number of British and American planes equally determined to prevent the Luftwaffe from getting at the ships.

The Germans relied heavily on their versatile twin-engined Ju88 Junker, which they nicknamed 'The Maid of All Work'. The Ju88 could perform the functions of a dive-bomber, a night fighter, a torpedo bomber or a heavy fighter. Her standard bomb load was around 4,000 pounds.

SS *Ocean Strength* and her convoy took evasion tactics by the procedure known as zigzagging, changing course every half-hour to confuse enemy submarines that may have detected her presence through their sonar.

The defensive air escorts did their damnedest to maintain a meaningful distance between *Ocean Strength* and potential attackers. From time to time, Supermarine Seafires – the naval equivalent of Spitfires – provided air cover.

As September 1943 drew to a close, and once again peering through his binoculars, Doug was able to witness an aerial combat taking place at least half a dozen miles away. A squadron of Heinkel torpedo bombers appeared at wave-top level and started to attack the merchant ships. Suddenly, Supermarine Seafires flying in from carriers such as HMS *Furious* and others dived down from 15,000 feet, swooped up behind the Heinkels and opened fire with their tracers and cannons. Doug saw the German bombers turn tail and head for their base in France. One of the bombers had smoke streaming from its starboard engine. It would be lucky to get back in one piece.

Five miles northeast of Christchurch in the rural English countryside of Hampshire, an aerodrome, near the village of Holmsley, was shared with the Americans and given the name RAF Station Holmsley South. It had grown from a hastily constructed airstrip in 1942 into a base accommodating 3,000 personnel with full amenities, communal areas and sick quarters. The ground-support station was constructed of Nissen and Maycrete huts where the mess facilities, chapel, hospital, armoury, parachute rigging, motor pool and other vital support functions were housed.

502 and 58 Squadrons were equipped with Handley Page Halifax Mark IIs whose chief task was an anti-shipping role. The Halifax, a four-engined heavy bomber, had an almost rectangular shape with a futuristic Perspex bubble for a nose,

inside which sat the front gunner with a Vickers .303 K gun. Two Merlin XX engines were slung under each wing. The rear gunner's turret poked up from the fuselage just above and behind the wings. The radar equipment was located in a cupola underneath the rear fuselage of the aircraft.

But it was a Ventura of the US Navy's newly arrived Iceland-based squadron that became the convoy's next saviour. Otto Finke, aged 28, was commander of the new U-279. He left Iceland on 25 September to sail south to join Wolfpack Rossbach. Piloting the Ventura, Charles L. Westhofn found a U-boat, U-305, on the surface and forced her to crash dive. Westhofn flew away to trick the U-boat into believing he had returned to base. However, as he started to return, he sighted a second U-boat, U-279, and flew against the flak to drop depth charges. Westhofn circled and had the satisfaction of seeing Finke's crew abandon ship and the U-boat turn turtle and plunge towards the bottom before he had to break off and return to base. Nothing more was heard of U-279.

On Monday, 27 September, the B flight of 58 Squadron were preparing for a U-boat patrol of the Bay of Biscay. The lead aircraft had a crew of eight, with Flying Officer Eric L. Hartley as Captain and Group Captain R.C. Mead as second pilot. Group Captain Mead was the station commander with a heavy workload, which prevented him from participating in as many flying missions as he could have wished. A Royal Canadian Air Force man, Flying Officer T.E. Bach was navigator, Sergeant G.R. Robertson was engineer and Sergeant Arthur S. Fox was wireless operator. The gunners were Flight Sergeant K.E. Ladds (mid-upper gunner), Sergeant R.K. Triggol (rear gunner) and Sergeant M. Griffiths (front gunner).

At the pre-mission briefing, the crews of 58 Squadron were told Bomber Command had sent orders that the codeword to contact the task force on this mission would be 'Seadog'.

The crews donned their fur-lined boots, their flying jackets and their helmets and prepared for a long flight.

'Contact!'

Flying Officer Eric Hartley's Halifax took off at 1128 hours. There was extra Avgas in the spare fuel tanks of the wing bomb bays to give flying endurance of up to 13 hours. They were equipped with the latest Mark III radar, Gee navigation and the Mark XIV bombsight. They carried eight 250lb depth charges.

By 1705 hours, B/58 had reached the limit of its patrol, and was returning to base from the western approaches. It was still daylight and the patrol had spotted nothing.

Flying Office Hartley gave the order. 'Seadog. Seadog. B Squadron will turn back and rendezvous over Land's End. Over and out.'

The other five Halifaxes under his command dutifully headed for England.

Eric Hartley turned to his co-pilot, Group Captain Mead, who happened to be his superior officer on the ground, and said, 'Would you like to take her the rest of the way, sir?'

'Thanks, Eric.'

At the very moment he was about to hand over the controls, Flight Officer Bach said into his R/T, 'Hey, Skipper, we've got a bandit on the radar – on the extreme edge of our range. Port side.'

Eric Hartley acknowledged his navigator: 'Roger.' He turned to his second pilot with a frown and said, 'I'm sorry, sir. Perhaps I should take a look.'

'By all means. Don't hang about though, Skipper. We've barely got enough juice to get back home.'

So, instead of following his squadron, Eric Hartley decided to investigate the possible sighting. He throttled back and flew lower.

A few minutes later, the convoy surrounding *Ocean Strength* became visible on the distant horizon. Dogging Convoy MKS 025 was U-boat U-221.

Ocean Strength was on his starboard side and U-boat Kapitanleutnant Hans-Hartwig Trojer was so intent on keeping out of radar range of the convoy that he did not at first see the lone Halifax approaching from his port side. A gunner on the aft deck was the first to spot the plane.

'Enemy on the port astern,' shouted the gunner.

The U-boat's engineer emerged from the conning tower. 'Come down, Herr Kapitan. We can crash dive.'

Hans-Hartwig saw the speed of the plane's approach. It would be above his U-boat in less than a minute. His reaction was automatic. 'No time. Stay topside. Fire all flak guns at will.'

Gunners scrambled to their positions and, within seconds, the approaching Halifax faced a barrage of cannon fire.

Hartley dropped to 50 feet, crossed the U-boat at 30 degrees and gave the order: 'Bomb-aimer, bomb-aimer ... Take a visual reckoning. When you're ready, release depth charges – minimum depth.'

'Roger.'

At 1713 hours, Sergeant Robinson triggered the mechanism release and eight depth charges fell on U-221 in quick succession.

'Bombs away!'

Seven charges of Amatol dropped into the water, surrounding the submarine and almost immediately exploded throwing up enormous plumes of water. They would have made life intolerable for the occupants of the sub, but, apart from leakages, which could be repaired, there was no mortal damage.

Then the final charge fell from the aircraft and landed with a dull thud on the conning tower. It didn't roll into the sea. In fact, it hardly moved at all. It appeared to be stuck there. The submarine rolled in the swell of the ocean. The waves buffeted against the depth charge.

Hans-Hartwig climbed a couple of steps and, using brute force, he tried to dislodge it. He was too late. The charge exploded and the man they called Count Dracula was blown to Kingdom Come. A gaping hole appeared along the upper superstructure of the sub and blackened faces began appearing from the mangled machinery below. There was no hope of survival. Mercilessly, the water rushed in. Incredibly, the U-221 continued to sail on while sinking gradually lower in the water. The Halifax, smoke pouring from one wing, circled to witness the sub's fate.

Several miles away, Chief Officer Henderson and Major Douglas Lidderdale stood on the bridge of *Ocean Strength*, binoculars in hand.

'There's so much bloody smoke,' said Doug. 'It's impossible to tell exactly what's going on. Surely that must have finished off the sub. It's got to sink now, wouldn't you say?'

'Look at the plane,' said Henderson. 'It's been hit.'

'What!'

Smoke trailed out of the starboard wing of the Halifax and began to engulf the fuselage. They watched helplessly as the plane continued flying away from them, maintaining a height of maybe 50 feet above sea level. Through their binoculars they thought they could see flames licking along the wing and enveloping the aircraft. The Halifax continued to fly, however, and disappeared over the horizon.

Meanwhile, the sub's bow rose to an angle of 20 degrees, then slid underneath the waves.

The convoy did not slow. No escort vessel turned back to search for survivors. There was nothing they could do, for there could be no survivors, just burning fuel oil floating on the water, a vast inferno like the inner circle of hell. And, in that carpet of flame, blackened bodies succumbed to their slow and torturous deaths.

Captain Rickard had managed to get on to the bridge. He looked thin and ill. He turned to his Chief Officer. 'Did we hear anything from the plane?'

'Nothing, sir.'

'It was on fire. Could they have put it out somehow?'

'Not unless they were able to gain height and go into a fast enough dive to extinguish the flames. The last we saw of them, they were limping along just above the surface.'

'We probably owe our lives to that plane.'

'Very likely, sir.'

'They might have ditched by now. Let's hope some other ship will come across them.'

Despite two of the escorts searching in a grid system for six hours, there was no sign of the Halifax. It could have limped along at sea level for some considerable distance, but nobody knew its precise fate.

Later that night, an officer came out of the communications room and handed the Captain a piece of paper. 'Skipper. Intelligence has been on the blower, sir.'

Captain Rickard read the communiqué and decided to take the Chief Officer and Major Lidderdale into his confidence.

'Ultra has intercepted another message from Admiral Doenitz. A new Wolfpack has been formed, code-named "Rossbach", to be positioned in a patrol line from naval square AK2645 to AK6817.'

'Forgive my ignorance, Captain,' said Doug. 'But where are those naval squares?'

The old man looked deathly pale as he replied. 'They are to

the north of us. But it's a loose arrangement. Doenitz is looking for convoys. And he is probably looking for MKS 025 in particular.'

Captain Rickard was correct. Fortunately, the Admiralty had the foresight to anticipate Doenitz's intentions too. They diverted Convoy HX-259 to the south and sent four destroyers of British Support Group 10 to reinforce the nine escorts of Canadian Escort Group C-2, and 39 merchant ships of convoy SC-143, using them as bait to lure the new 'Rossbach' group and hopefully wipe it out. The Sea Lords reasoned that these diversionary tactics would provide an opportunity for *Ocean Strength* and her escorts to make a clear run for Scotland.

With dusk came a lessening of tension. Doug and his companions started to relax. The time of maximum danger was the dawn. As the sun rose, so did the U-boats.

That evening, they ate in the mess – an excellent meal of corned beef and rehydrated potatoes prepared by the two Singhalese cooks in the galley, specialists in the art of tinned cuisine.

The Halifax had not survived. Sergeant Arthur Fox sent out frantic SOS calls but he feared his radio equipment had been compromised. Flying Officer Eric Hartley ditched the bomber at a speed of 110 knots without flap. The tail broke off, water rushed in and the aircraft stood on its nose. The fire went out.

Sergeant Triggol, the rear gunner, crawled out of his smashed turret, but slipped on the fuselage and disappeared into the sea. He was never seen again.

Sergeant Griffiths, the front gunner, managed to extricate himself from his Perspex bubble and reached the cockpit. He had been hit and was bleeding heavily. Eric Hartley held him.

'We are getting the dinghies out. You'll be all right, old fellow,' he said hopefully.

But Sergeant Griffiths had lost too much blood and was already unconscious. Group Captain Mead and Flying Officer Hartley abandoned the aircraft carrying Griffiths' dead weight between them, but a sudden wave forced him from their grasp and he slipped away into the sea. There was nothing they could do to save him. The crew had managed to jettison only one of the dinghies and the six survivors struggled to inflate it, while at the same time trying to put as much distance as possible between themselves and the vortices of the sinking aircraft.

Their jackets and flying boots would probably be of great use to them in the future but, for the moment, the sodden clothing only hampered them. Fortunately, they were all wearing Mae Wests, which probably kept them alive. Eventually, they succeeded in their mission and wearily clambered into the rubber boat.

They paddled as far from the crippled Halifax as they could and watched it nose-dive slowly into the deep as darkness fell.

The convoy and its escorts did not relax their vigilance. There was a constant lookout and many eyes strained over the ASDICs – the sonar device that was the only safeguard against U-boats. It could pick up the signs of a sub up to two miles away.

Radar scanners attached high on the masts swept ceaselessly in a 360-degree arc. Down in the radar room, relays of operators watched the glowing screens and could detect surface ships and airplanes up to 50 miles distant, but they found nothing of the plane or the dinghy.

CHAPTER THIRTY-SEVEN
SEPTEMBER 1943

U-boats were at their most dangerous first thing in the morning. Their standard procedure was to approach a convoy from the northwest at dawn. This allowed them to have their targets silhouetted against the sunrise.

The morning of Friday, 28 September found Flying Officer Hartley and the other five survivors of the Halifax huddled in their dinghy, cold, wet, hungry and totally alone. Flight Sergeant Ladds and Sergeant Fox were both seasick and spent a great deal of time hanging over the edge of the dinghy, studying in detail the serried waves of icy midnight-blue, each wave capped like a patisserie cake with frothing cups of cream. This sight did nothing to relieve them of the churning in their stomachs.

Eric Hartley, meanwhile, had discovered the emergency rations, which consisted of Horlicks tablets, barley sugar, chewing gum, chocolate, condensed milk, salt tablets and a packet of three condoms. He held these out to his companions.

'I've heard of shipwreck victims resorting to cannibalism, and I can just about grasp the concept,' he said wryly. 'But there is one other thing, chaps, which I'm afraid I refuse to countenance.'

Group Captain Mead explained. 'Emergency packs are nearly all the same. The thinking behind rubber condoms is, I understand, that, if you're forced to ditch in the jungle or in the Arctic among Eskimos, the local chief may offer you one of his wives to sleep with. Well, it might be taken as an insult to say no and so rather than create offence ... as a precaution...'

There was silence for a while. It was difficult to achieve levity under the circumstances.

'Arthur, did you have time to send off the SOS signal?' said Hartley at length.

'I tried to, sir,' replied Sergeant Arthur Fox, the wireless operator. 'Trouble is, I have no way of knowing if the equipment was knocked out. I did the standard May Day procedure, but we are several hundred miles from England and we were only flying at about 300 feet. I don't know if the signal would have carried.'

'All we can do is pray that it was picked up. Part of the emergency rations includes five pints of water. That may be the difference between survival and death for us all. When it rains – and I promise you that it is bound to bloody rain – I suggest we all try to catch as much water as we can and try to replenish these five containers. For the first 24 hours, we will not touch anything. We can live off our own body fat for a while. If any of you fishes as a hobby, this would be a jolly good time to work out how you might catch something here. Keep as warm as you can and keep your eyes peeled. We have four Very light cartridges to attract attention and we don't want to waste them. Now I'm going to conserve energy and shut up for a bit. And I strongly urge you all to do the same.'

Flying Officer Hartley pulled down his flying goggles – Halcyon BS 4110 – and shut his eyes.

The others looked at each other glumly. They were well wrapped up in their Irvine jacket flying suits and helmets, but each man realised that his prospects for survival were desperately poor.

The following day was calm and sunny. Sergeants Ladds and Fox had recovered somewhat from their discomfort. Flying Officer Bach, a rugged Canadian, stripped and swam for ten minutes. The appearance of jellyfish prompted him to take up residence in the dinghy again. Each man was allocated a barley sugar and a square of chocolate. By the end of the afternoon, they were down to two pints of water. Sergeant Robertson had improvised a fishing line with a safety pin which he hung over the side but all he caught were two jellyfish. He tried to dehydrate them in the sun and then cautiously tasted one. He spat it out in disgust. He said it tasted very bitter.

Group Captain Mead had the men take it in turns with the two surviving paddles to aim in a northeasterly direction. He reasoned that, if they could position themselves far enough north, with luck they might find themselves carried on the current of the Gulf Stream. Instead, the weather grew colder and the sea rougher. Squalls were ahead.

Two more nights passed. Nothing had been caught. Nothing had been sighted. The men made each barley sugar sweet last as long as possible. They huddled for warmth. They were becoming dehydrated. They took it in turns to keep lookout. One night, Sergeant Robertson woke up the sleeping men with a hoarse cry.

'I see a light,' he cried. 'Look – up there. On the horizon. It's a plane. I swear it's the Leigh light of a Liberator or a Wimpey. Where is the Very pistol? Give it to me quickly...'

Sergeant Robertson fired a one-inch-bore Very cartridge into the air. The brightness of the starburst dazzled them. The light on the horizon stayed obstinately still.

Group Captain Mead stared at the distant orb of light. 'It's rather reddish for a Leigh light,' he said reflectively.

Robertson, anxious for his sighting not to prove a false hope, said, 'We can't properly judge from this distance. Must be three or four miles away.'

'Why would he switch on his Leigh light unless he'd spotted a submarine?' said Hartley.

Flight Sergeant Ladds said, 'There's no sign of ack-ack or depth charges.'

Finally, Flying Officer Bach started chuckling. 'What damn fools we all are. That's not a plane. It's a planet.'

'What are you saying?' said Robertson.

'I'm saying that's the red planet. It's shining quite bright tonight. We've just sent up a Very light to signal Mars.'

By morning, the sea had whipped itself into a frenzy and the wind became paralysingly cold. Clinging on to lanyards and each other, the six men slid into a heap and the dinghy overturned. Part of their training course involved practising in swimming pools how to right-side an upturned dinghy. After half an hour of extreme physical labour and discomfort, the training paid dividends and they lay exhausted in the little craft, praying for sufficient strength to resume bailing out water.

It rained the following day. They tried collecting the rain in their handkerchiefs and transferring the water into the empty bottles, and managed to make three pints. Arthur Fox, the wireless operator, slipped into semi-consciousness and began mumbling deliriously.

The waves became rougher. The dinghy shipped water. The men spent all their time bailing out the water with their flying helmets.

As the Convoy MKS 025 finally sailed into the relatively safe waters of the Clyde river, the six survivors from the Halifax were still at sea, huddled in their rubber dinghy, barely conscious and with their last hopes of rescue vanishing.

The days passed. October arrived. The spiritual reserves of the survivors waned to a low ebb. The situation seemed hopeless. They managed to revive Arthur Fox with the last of the condensed milk. Then they had nothing left. Weakened mentally and physically, they all began showing the symptoms of dehydration and delirium.

On 8 October, after 11 days at sea, at 1400 hours, a ship was sighted on the distant horizon. It was impossible to tell whether it was friend or foe. By this time, nothing mattered but getting out of the leaking dinghy. Eric Hartley, so mentally and physically numbed he barely knew what he was doing, fired the remaining three Very cartridges. At 1500 hours, three destroyers, *Mahratta*, *Matchless* and *Valiant* arrived and hauled the exhausted men to safety.

They were taken to Plymouth, full speed ahead. Two days later, they were resurrected in the Royal Naval Hospital.

It transpired that HMS *Mahratta*, which had been bound for Gibraltar, came across the dinghy by sheer chance. No SOS had ever been picked up.

CHAPTER THIRTY-EIGHT
OCTOBER 1943

On the same day that Flying Officer Hartley's surviving crew were being rescued, SS *Ocean Strength* found sanctuary in her final destination, Glasgow. Major Douglas Lidderdale disembarked and made a beeline for the nearest telephone. The first call he made was to his beloved bride from whom he had been forcibly separated for so long.

His second call was to the Mechanical Engineering Directorate at the War Office where he spoke to one of Major-General E.B. Rowcroft's aides-de-camp. REME had been formed the previous year to recover and maintain the army's equipment both in the battlefield and in the rear. Eric Bertram Rowcroft remained its first Director.

It was arranged for a top-security dispatch rider to rendezvous with Doug within the next 24 hours.

Doug had completed his report on Tiger 131's specifications the day before he had left Bone. But he had not finished typing up his top-secret mission report until home waters were within sight. He was, by nature, naturally reticent, and he wrote in an oddly detached manner.

Later, he stood on the quayside and watched as the sailors off-loaded ten crates of spares and those Tiger tank body parts that had been scavenged from other wrecked Tigers during the first part of their mission.

Thus far, he had faithfully fulfilled every particular of Winston Churchill's original directive. His mission report should be in the Prime Minister's hands within 36 hours. Now Doug pondered the final challenge. How was he to display Tiger 131 to the world outside the Prime Minister's headquarters in Downing Street?

James Morrison Henderson came out of the offices of the Ship Owners Denholm and Company, and walked across the cobbles to join Doug. He appeared to have a spring in his step.

Doug gave his new friend a smile. 'Thanks for everything, Chief. You got us here in one piece. How is Captain Rickard?'

'He is due for a well-earned rest. He's had one helluva war.'

'I've never heard him complain.'

'Exactly. A good man. I shall miss him.'

'And where are you off to next?'

'Russia. But I stay with *Ocean Strength*,' Henderson said as casually as a proud man could. 'As a matter of fact, I am the ship's new Master.'

'Congratulations. This calls for a celebratory cocktail.'

'Later, Doug. Let's see that the Tiger makes it safely to shore.'

They watched as Driver Wilkes drove Tiger 131 from the foredeck of the Liberty Ship and on to the quayside. Dockers came to a standstill to watch the lumbering beast.

'With her tendency to overheat, it'll be chancy to drive her the whole way, but we can if you want to,' said Pumfrey.

Doug grinned. 'Good God, no. I can't be expected to keep on repairing the Germans' slip-shod engineering. From now on, she shall be piggybacked everywhere. Besides, she burns so much fuel, at 2.75 gallons per mile, she'd bankrupt the country.'

'So which of us is having the first turn to give her a piggyback?'

'I've phoned Pickford's.'

'Pickford's! You mean the furniture-removal company?'

'The very same. The army doesn't have any 60-ton tanks of its own and certainly nothing available up here that would lift much more than a Churchill.'

'Whereas Pickford's, because of all those heavy grand pianos, does?'

Doug laughed. 'Exactly. Those and some hefty ship turbines and prefabricated bridge parts. I'm letting Pickford's take the strain with a 100-ton Scammell transporter. All you have to do is get our Tiger on to it.'

'Are you coming too, sir?'

'I'll join you later. See you in Tothan on the 20th for the tests. Right now I have a train to catch.'

Pumfrey grinned. 'I hope I'm not overstepping the mark if you'll allow me to wish you good luck, sir.'

'Eh?'

'I remember you saying you hadn't had time for a honeymoon.'

Doug sighed. 'That comes later. First, I have to finish some official business.'

Pumfrey saluted and Doug responded before hurrying away to the railway station.

Major Aubrey Douglas Lidderdale had been summoned to the War Cabinet Rooms under Whitehall. Accompanied by Leading Wren Driver Mackintosh, who possessed all the strict security clearances, he descended the spiral staircase to the Minoan-like maze below. Here he was greeted like a long-lost friend by Joan Bright who escorted him to Room 13.

John Masterman was chairman of Section 17M, otherwise referred to as the Naval Intelligence Department's (NID) 'Twenty Committee'. With him was Ian Fleming, who would later become a novelist and invent the super spy James Bond. Behind a desk sat Rear Admiral Edmund Rushbrooke, who had just been appointed as the new head of the NID. He sat and watched and said nothing throughout the proceedings. The NID did not confine its covert activities solely to naval matters.

Major Desmond Morton opened the door and courteously ushered in Douglas Lidderdale. This was the first top-level debriefing that Doug had ever attended and he did not know what to expect.

'Welcome back, Major,' said Morton. 'And congratulations.'

'Thank you, sir.'

'Lidderdale, let me introduce you to Mr Masterman and Commander Fleming. The intelligence service, you know.'

Doug suspected he did know. These men were senior men in the secret services about which the public at large heard only whispered rumours.

Masterman shook Doug's hand warmly. 'Mr Churchill conveys his apologies. He would have liked to congratulate you personally, but he is heavily engaged elsewhere at the moment.'

'I quite understand.'

'Major Morton relayed to me your dealings with one of our operatives in Algeria. A Mr, er, Craker. Is that correct?'

'Yes, sir.'

'You are quite sure his name is "Craker"?'

'Yes, sir. But Brigadier Tetley referred to him as Colonel Clarke. He lent an ear, gave advice and proved most useful. Actually, I was wondering whether you might be able to get a message to him conveying my appreciation for all his help.'

Doug looked round at the other men in the room who were

observing him in silence. He was conscious that the smile on his face seemed a little forced. There was an awkward lull in the proceedings.

Desmond Morton, whose job included having a foot squarely in the middle of MI6, offered Doug a cigarette. He accepted gratefully and they both inhaled with some relief.

Then John Masterman broke the ice. 'Remind me, Major, how did Mr Craker reveal to you that he is one of our agents?'

'He never claimed to be an agent, sir. I just assumed him to be. Dudley Wrangel Craker.'

Ian Fleming exhaled a cloud of smoke. 'Did he exhibit any effeminate tendencies?' he said softly.

The query took Doug aback. He hesitated before replying. Then he chose his words carefully. 'Well, not in my presence, sir.'

'He didn't dress up as a woman, then?'

Doug looked at Ian Fleming in astonishment. 'I never saw him dress up, sir.'

Fleming moved a step closer to Doug and looked into his clear, cool, grey-blue eyes – the very description he would one day apply to his fictional hero James Bond.

Masterman spoke. 'I'm sorry to inform you that none of our agencies has any record of a Dudley Craker.'

'Not even in Special Services, sir?'

'We *are* the Special Services, Major. And, on reflection, the name is rather improbable, don't you think? Wrangel Craker. Sounds somewhat contrived, don't you think?'

Doug frowned. 'He was eccentric, certainly, but he seemed so genned up. Everything he predicted actually came to pass. And, if you don't mind me saying so, Dudley Wrangel Clarke sounds equally improbable.'

Ian Fleming chuckled. 'Quite right, Major. It does. But Mr

Clarke does exist. He is an occasional war correspondent for *The Times*.'

'We must all be on our guard, Major,' said Masterman. 'The man may have been genuine but, for some security reason or another, decided to use a pseudonym. Who can tell? The safest thing in wartime is to assume that the enemy is at our gates. And that the Trojan horse may have spies hidden in it. All in all, it's not worthwhile taking chances.'

'It might be best to forget about this person, Lidderdale,' said Fleming. 'And, if your exploits with the Tiger ever come up, it may be as well to tiptoe away from the subject.'

Doug nodded. 'I quite understand, sir.'

'Well, you've successfully accomplished your mission. The Tiger is safely on our shores. Congratulations, Major.'

Mr Morton rose to go. 'Would you excuse me? I have a meeting to attend.'

'Thank you for your time, sir. I'd be obliged if you'd extend my compliments to Mr Churchill.'

'Certainly, certainly.' He shook hands with Doug and gave the impression that he had other urgent things to attend to.

Doug was on the verge of leaving Room 13 when the door was pushed open with some vigour.

A familiar voice bellowed, 'Mr Morton? Where are you?'

Winston Churchill, a cigar clenched in the left side of his mouth, entered and looked round. At the sight of Douglas Lidderdale, he strode forward and offered his hand, which Doug happily accepted.

'You made it back, Colonel. I've read your reports. Excellent, excellent.'

'Thank you, sir.'

Churchill's ebullience was rather overwhelming. 'When you've stripped her down, figure out how to put all the Tiger's tricky bits into our own machines. Give our tanks something of the same firepower. Then, I want her displayed for the nation; a physical symbol of our impending victory. Mr Morton, arrange for it to be erected at the back of Number Ten for the whole world to see.'

'Yes, sir.'

'Mr Schicklegruber and his Huns will be crying themselves to sleep tonight. Thank you, Colonel.'

Doug beamed with pleasure at the great man's effusiveness. 'Thank you, sir. But, if I may say so, sir, with respect, I am only a Major, sir.'

Churchill's large face wrinkled up with genuine amusement. 'With respect, Colonel, I think you are in error.'

'Whatever you say, sir.'

'I think you can call yourself Colonel after this, without fear of contradiction. Well done again.'

Winston Churchill hurried out of the room followed closely by his private secretary.

Doug watched the Prime Minister disappear in the direction of his own rooms.

'Come along, Colonel,' said Morton pointedly. 'Allow me to guide you out of this maze.'

Lieutenant Colonel Aubrey Douglas Lidderdale stubbed out his cigarette on one of the many ashtrays lining the corridor as he followed, his forehead furrowed in thought.

Messrs Fleming and Masterman had been somewhat disingenuous regarding Dudley. They knew him and his *noms de guerre* perfectly well. He was one of the few agents to have been given a free hand by Winston Churchill.

Doug and Kathleen had prearranged that their reunion should

take place outside Madame Tussauds famous wax museum in Marylebone Road. It had stood on its present location for 60 years and the chances were it would remain intact for the duration of the war.

The unlit streets at night made driving and walking difficult. Pavements were like obstacle courses. The few pedestrians that ventured out flashed an occasional torch to avoid sandbags and broken glass.

Lieutenant Colonel Douglas Lidderdale came out of Baker Street tube station and hurried as fast as he could through the debris until he saw the Model T, with its engine running, parked outside Madame Tussauds. He stubbed out his cigarette, opened the passenger door, slid into the car and, without saying a word, embraced his bride. There were nine months of kissing to catch up with. Time passed.

They were interrupted in their deliberations by a police constable knocking on a side window. They could only just make out what he said. 'You can't park here, you know. You're not one of the exhibits.'

Doug called back amiably, 'Sorry, officer. Just getting warmed up.'

'Yes,' replied the policeman, 'I can see that.'

Doug said to Kate, 'Move over. I'm driving.'

He got out of the car and went round to the driving seat, by which time Kate was in the passenger seat looking puzzled.

'Why did you want me to bring the car?' she asked. 'I told you I've rented an apartment in Rossmore Mansions. It's only a few minutes' walk away.'

Doug released the hand brake and started to motor towards Euston.

Kathleen looked concerned. 'Rossmore is the other way,' she said. 'I've put down a deposit and gone to a lot of trouble to make it all nice and cosy for us ... It was going to be our nest.'

'First things first,' said Doug with an air of mystery.

He drove along the Old Kent Road and continued in a southeasterly direction. A couple of hours later, they entered the outskirts of Folkestone by which time Kathleen's demeanour had become markedly less sunny. This was not how she had anticipated her new husband's homecoming after nine months' absence.

Eventually, Doug braked and turned into the gravelled driveway of a half-timbered hotel called The Tudor Tower.

'I haven't even got an overnight bag with me,' said Kate.

'You won't need anything,' replied Doug.

'What on earth are you up to?'

'I'm finishing our honeymoon.'

Kathleen gave a peel of laughter.

'Don't laugh. There's more to marriage than signing a piece of paper.'

Later that night, their hotel room was lit by a single candle shoved into the mouth of an empty champagne bottle.

Her face lay on the pillow next to his.

It was nearly dawn. 'What adventures did you have?' she whispered.

Doug smiled sleepily. 'I'm under orders to tiptoe away from the subject,' he said.

And they both went to sleep, cosy at last, in each other's arms.

EPILOGUE

In November 1943, still under Doug's command, Tiger 131 was paraded on Horse Guards Parade, where it was offered as a present from General Anderson, Commander of 1st Army, to Winston Churchill, who was very ill at the time in Morocco. The cannon of the tank was pointed into the gardens of 10 Downing Street.

Doug went on to be responsible for the special armoured assault equipment used in the D-Day invasion. Under the general command of Major-General Percy Hobart, Doug was in charge of the Special Devices Branch of DTD. He designed flails and flame-throwing devices based on some of the technical discoveries gleaned from the Tiger 131. These were deployed to clear the beaches of mines and to enable landing infantry and armoured equipment to have a clear run into Normandy.

Winston Churchill wrote in his memoirs (*The Second World War*, Volume 5): 'Space forbids description of the many contrivances devised to overcome the formidable obstacles and minefields guarding the beaches. Some were fitted to our tanks to protect their crews; others served the landing craft. All these matters aroused my personal interest, and, when it seemed necessary, my intervention.'

On 21 September 1943, it was announced in the *London Gazette* that the Military Cross (for bravery) was awarded to Lieutenant Reginald Thomas Whatley (231260), Royal Electrical and Mechanical Engineers (Totland Bay, I.O.W.). Whatley's citation for the MC reads as follows:

> During the fighting between 29 April – 6 May, a number of Churchill Tanks became casualties in the neighbourhood of Pt. 174, Medjez-el-Bab sector. Lt. Whatley displayed great coolness and determination in going out to recover those tanks, most of which were in no-mans-land. He was so successful that with 2 NCOs he was able to recover and bring back 12 Churchill Tanks to our own lines. The damaged tanks were under full observation of the enemy, and consequently the use of a recovery vehicle immediately drew enemy mortar fire.
>
> Despite this, Lt. Whatley performed his duties with remarkable energy and was constantly under shell and accurate mortar fire.
>
> Most of the Churchills were in the midst of minefields strewn with Teller and anti-personnel mines, but this did not deter Lt. Whatley from his purpose.
>
> His complete disregard for his own safety and his devotion to duty throughout were an example to all. If it had not been for his tenacity the recovery of these 12 tank might never have been achieved.

24 May '43: typed.

Signed – In The Field – 31 May 1943 – [with a handwritten recommendation] by Lieutenant-General Charles Walter Affrey, Commander of 5 Corps. The Military Cross was actually awarded on 29 September 1943.

RAF pilot Joseph Berry, whom Doug befriended in Algiers, flew continuous missions in North Africa, Sicily and Italy before returning to England as Squadron Leader, where he became an ace flying bomb killer with 60 destroyed V-1 and V-2 rockets to his credit. His plane was hit by enemy fire during a mission over Holland in November 1944 and crashed in flames. He had been awarded the DFC and Bar. He was 24.

U-boat 371, at which Doug fired the Tiger's gun from the SS *Empire Candida* in August 1943, was sunk by a force of Allied destroyers in the Mediterranean in May 1944. Kapitanleutnant Waldemar Mehl went on to command two other U-boats until the end of the war. He was awarded the Knights Cross in March 1944. Mehl died in Germany in 1996 aged 81.

After the war, Brigadier James Noel Tetley DSO, TD, DL, LLD – known as Noel – became an ADC to King George VI and to the Queen before rejoining his family's brewery business in Leeds. He died in 1971 aged 72.

Brigadier Richard Maxwell CB also became an ADC to the Queen. He had been severely injured by a land mine in Tunisia in May 1943 and stretchered back to England. He became a Provost Marshal before retiring from the army in 1955. He died in 1965 aged 66.

Major Peter Gudgin remained with the Royal Tank Regiment until 1969, after which he became a noted military historian. His books include *Military Intelligence: A History*, and *With Churchills to War*, his superb history of the 48th Battalion RTR. He also wrote the foreword to the Bovington Tank Museum's celebrated manual on the Tiger tank, published in 2011. Major Gudgin's long correspondence with Colonel Lidderdale about Tiger 131 was kindly made available to the authors by the Museum Archives. Major Gudgin died in September 2011 aged 88.

It was announced in the *London Gazette* on 14 October 1943 that Lieutenant Colonel Dudley Wrangel Clarke of the Royal Artillery (Service Number 13136) had been made a Commander of the Most Excellent Order of the British Empire. Over 30 years later, documents were released to the public that disclosed how Dudley had been instrumental in creating Section MO9 of the War Office on 8 June 1940. Thus, the Commandos were born. He had part of an ear shot off in the very first Commando raid in the summer of 1940, after which Dudley left his house in Stratton Street, Mayfair, and travelled from one theatre of war to another. Winston S. Churchill had given Dudley a free hand to organise lightning commando strikes on his own volition. His younger brother T.E.B. Clarke was a writer who won an Oscar for his film *The Lavender Hill Mob*. Other works included *Passport to Pimlico* and *The Titfield Thunderbolt*.

'Quadrant' was the codename for the Quebec conference, being a series of meetings between President Roosevelt, Winston Churchill and their Security Chiefs. Churchill, under the assumed name of 'Colonel Warden', travelled to the conference on the *Queen Mary* accompanied by his family.

Joan Bright was the head of the Secretariat that looked after the Interservices Security Bureau, part of MO9, in the underground Cabinet War Rooms. She had a brief fling with Ian Fleming and is considered to be the model for Miss Moneypenny in the James Bond novels.

Rear Admiral Edmund Rushbrooke's predecessor as Head of the Naval Intelligence Department was Admiral John Godfrey CB, who had been Ian Fleming's mentor. Godfrey was replaced on Churchill's personal orders. Godfrey was the model for 'M' in the James Bond stories.

The 30th Assault Unit was formed under the wing of the Naval Intelligence Department by Ian Fleming. Members of the unit included Ralph Izzard and Patrick Dalzel-Job. They were influential in the creation of the character of James Bond.

On 12 March 1945, Kathleen Lidderdale had a fine baby boy called David Anthony who in later life changed his name to David Travis and became a highly successful singer and record producer.

Ten years after the end of the Second World War, Douglas Lidderdale had become chief engineer of Telcon and Submarine Cables Ltd, responsible for setting up a complete new plant for the manufacture of the first transatlantic telephone cable and for re-equipping the cable ship *Monarch* to lay it. He retired in 1984.

Captain Rickard died at the age of 58 but, much earlier, James Henderson had taken over as Captain of SS *Ocean Strength*. He and Doug continued to indulge in a lively correspondence for many years. Henderson was heavily engaged in the top-secret pipe-laying project called OPERATION PLUTO (Pipe Line Under The Ocean), which carried fuel from Britain to the Continent. At the time, this was referred to as 'cable laying', so that no inadvertent reference would reveal that a pipe-laying scheme was happening. Each mile of pipe used 24 tons of lead, 7.5 tons of steel tape and 15 tons of steel armour wire, eventually resulting in a million gallons of fuel per day being pumped across the channel.

By the end of the war, Douglas Lidderdale (serial number 122532) had been made up to a full Colonel. In 1946, upon demobilisation, Doug was made to drop a rank, and returned to being a Lieutenant Colonel. This allowed a grateful government to pay him a lower pension.

He died in June 1999. Kathleen predeceased him in 1995.

David Travis (Lidderdale) lives on and is the factual adviser for this book.